Analyzing
Informal Fallacies

Analyzing
Informal Fallacies

S. Morris Engel

School of Philosophy
University of Southern California

PRENTICE-HALL, INC., Englewood Cliffs, New Jersey 07632

Library of Congress Cataloging in Publication Data

ENGEL, S   MORRIS, (date)
   Analyzing informal fallacies.

   1. Fallacies (Logic)   I. Title.
BC175.E52   165   79–29643
ISBN   0–13–032854–5

Editorial/production supervision and interior design by Judy Brown
Cover design by Judith Winthrop
Manufacturing buyer: John Hall

Printed in the United States of America

10   9   8   7   6   5   4   3   2   1

Prentice-Hall International, Inc., *London*
Prentice-Hall of Australia Pty. Limited, *Sydney*
Prentice-Hall of Canada, Ltd., *Toronto*
Prentice-Hall of India Private Limited, *New Delhi*
Prentice-Hall of Japan, Inc., *Tokyo*
Prentice-Hall of Southeast Asia Pte. Ltd., *Singapore*
Whitehall Books Limited, *Wellington, New Zealand*

to Trudy

with affection

Contents

Preface to Students ix

Preface to Instructors xiii

PART ONE

**Fallacies
of Ambiguity** 1

Answers to Exercises 29
Test on the Fallacies
of Ambiguity 35
Answers to
Test on the Fallacies
of Ambiguity 38

PART TWO

**Fallacies
of Presumption** 39

Answers to Exercises 100
Test on the Fallacies
of Presumption 117
Answers to
Test on the Fallacies
of Presumption 125

PART THREE

Fallacies
of Relevance 127

Answers to Exercises 164
Test on the Fallacies
of Relevance 170
Answers to
Test on the Fallacies
of Relevance 173

Appendix 175

Preface
to Students

The study of common fallacies has been an integral part of the study of logic ever since the time of Aristotle, who was the first to investigate them with care and detail. Although subsequent investigators have not hesitated to depart from the distinctions and divisions Aristotle introduced, the majority of present-day writers seem to find the three major groups adopted here—fallacies of ambiguity, of presumption, and of relevance—the best arrangement in view of the material.

It is not clear who first proposed and used this threefold division of the fallacies, nor is there agreement about the distinctions meant by those who use these terms. Furthermore, although some types of fallacies seem more deceptive and serious than others, writers differ regarding which ones can be safely ignored. As a result, no two texts are likely to contain precisely the same number of fallacies.

There is, however, almost universal agreement regarding the importance of studying these common fallacies, even if this sometimes seems belied by the somewhat absurd and artificial examples various authors use to illustrate these logical errors. But even these absurd examples of reasoning have their place in this sort of study, serving the same function in logic as the telescope or microscope in other disciplines; they enlarge the nature of the difficulty being examined so we can see it more clearly and therefore learn to deal with it more effectively. This is why I have not hesitated to include in this collection a number of witticisms, even though these are not fallacies but merely exploitations of the same devices found in fallacies.

I have also tried to strike a balance in the examples collected (which have been gleaned from sources such as newspapers, books, political speeches, and letters to the editor) between briefer and less complex examples and lengthier and more confusing ones that might be encountered in discussions and debates on pressing issues. Before we can deal effectively with more complicated and somewhat more scattered arguments, we must develop skill in dealing with examples in which the different types of irrelevances, confusions, and difficulties that predictably arise in arguments stand out more clearly. Only after mastering these difficulties can we properly turn to the task of unraveling the longer, more involved, and more confusing articulations. The practice gained in dealing with the somewhat briefer and less complex examples, in other words, is similar to practicing musical scales. Although no one will ever be called upon to give a concert of scales, learning to play them enables an individual to play concert music. I hope these seemingly easier and somewhat more artificial cases will not be overlooked but will be used to gain this valuable skill.

No one has put this better than Gilbert Ryle:

> Fighting in battles is markedly unlike parade-ground drill. The best conducted drill-evolutions would be the worst possible battle-movements, and the most favorable terrain for a rearguard action would entirely forbid what the barrack-square is made for. None the less the efficient and resourceful fighter is also the well-drilled soldier. The ways in which he takes advantage of the irregularities of the ground show the marks of the schooling he had received on the asphalt. He can improvise operations in the dark and at the risk of his life now, partly because he had learned before to do highly stereotyped and formalized things in broad daylight and in conditions of unmitigated tedium. It is not the stereotyped motions of drill, but its standards of perfection of control which are transmitted from the parade-ground to the battlefield.[1]

And again a few pages later:

> To know how to go through completely stereotyped movements in artificial parade-ground conditions with perfect correctness is to have learned not indeed how to conduct oneself in battle but how rigorously to apply standards of soldierly efficiency even to unrehearsed actions and decisions in novel and nasty situations and in irregular and unfamiliar country.[2]

The division of the fallacies into the standard three groups followed here fits the subject very well. Logic is the study of argument; and before giving our assent to an argument, we should always make sure we are clear about the following: (1) Is what the argument asserts clear? (2) Are the facts in the argument correctly represented? (3) Is the reasoning in the argument valid? The three traditional categories are

---

[1] Gilbert Ryle, *Dilemmas* (Cambridge, England: Cambridge University Press, 1960), p. 112.

[2] Ibid., p. 123.

tied to these three aspects of arguments. Thus the first set (the fallacies of ambiguity) deals with arguments that fail to meet the challenge of the first question. The second (on the fallacies of presumption) deals with arguments that fail to meet the challenge of the second question. The third (on the fallacies of relevance) deals with arguments that fail to meet the challenge of the third question.

In dealing with these fallacies your goal should not be simply to identify the fallacy in question (that is, to tag it with some appropriate label) but to develop the skill of explaining clearly and precisely why this or that argument is less than sound. Let us take, for instance, the following typical example of the fallacy of amphiboly:

> The concert held in Good Templars' Hall was a great success. Special thanks are due the vicar's daughter, who labored the whole evening at the piano, which as usual, fell upon her.

Its analysis may take the following form:

> From the somewhat careless way the second sentence of the argument is constructed, it could be interpreted to mean either that (a) the piano itself fell on the vicar's daughter or (more likely) that (b) the *labor of playing it* fell on her. What confuses us here is the term "which," whose antecedent is somewhat ambiguous.

Normally, you will need only two or three such sentences to explain the example and expose its fallacy. The object is to develop the skill to do so with economy and clarity.

There are in this book some examples that have the appearance of statements rather than arguments and therefore seem to be out of place. These statements, however, are highly abbreviated arguments and are included because that is the way we all too often argue (since time is short and thought is quick, we omit much that seems unnecessary to state explicitly). Thus, pointing a finger at someone and shouting, "*Liar!*" can be considered an argument, for what is asserted (not that anyone would ever take the trouble to do so explicitly) is the following:

> All people who try to deceive others by uttering what they know to be false are liars.
>
> You are such a person.
>
> Therefore you are a liar.

Of course, establishing the full text of the intended argument (which is another important and useful skill to try to develop) does not make the argument sound; it does, however, place us in a better position to de-

termine whether it is or not. To reject such examples from consideration is to cut ourselves off from much actual discourse.

As an aid to self-study I have provided a brief explanation of each
of the fallacies as they have been traditionally divided among the three
major categories, answers to over a third of the problems, and a self-test
at the end of each section.

Good luck.

Preface
to Instructors

As anyone who has ever taught the informal fallacies has surely experienced, the greatest difficulty is sufficiently distinguishing between the various fallacies to enable students to apply them easily and confidently. Nevertheless, I have retained the distinctions because I have found using them, despite the borderline cases that elude easy classification, is both challenging and rewarding.

Each person, however, understands these traditional classifications differently. I shall therefore explain how I have come to understand these traditional fallacies, what unites the various fallacies generally placed in each of the three major traditional divisions, and what I have found valuable in teaching this material.

The fallacies I usually discuss under ambiguity are amphiboly, accent, hypostatization, equivocation, and division and composition. What tends to deceive us in the case of the fallacies of ambiguity is the confusing nature of the language in which the argument is expressed. Each type of confusion centers on an important aspect of the nature of sentences.

1. Amphiboly explores the consequences of not taking sufficient care in structuring sentences or statements.
2. Accent explores what can go wrong when we mistake the context of a sentence or statement and as a result fail to understand it in the way it was intended.
3. Hypostatization explores what is entailed when we mistake the reference of a sentence or statement as a result of confusing concrete terms with abstract ones.
4. Equivocation explores the errors we are prone to commit as a result of failing to recognize that many words have multiple meanings.
5. Division and composition explore the mistakes we make as a result of confusing the distributive and collective senses of terms.

Part One illustrates these six fallacies. From the answers, or hints of answers, given (which, if read consecutively, provide a fairly detailed account of the nature of the fallacy illustrated), I have tried to show why I categorize them as I do. But one person's amphiboly is often another's equivocation, and you may disagree with me. Such disagreement should cause neither discouragement nor distress, for the study of informal fallacies is not the study of mathematics or natural science. As Aristotle said, we should expect from a subject only that degree of precision of which it is capable. I tell my students we will inevitably run across cases that elude precise classification; it is sufficient if students can present a clear account of why they think an example is amphiboly rather than equivocation.

It will happen that stressing a certain feature in a given statement or argument (say, its odd sentence structure) will allow us to explain the resulting confusion by amphiboly, whereas stressing the apparently equivocal use of a certain word within that same example may allow an explanation of equivocation. In those cases where it proves easier to clarify the confusion in one way than the other, the easier way is obviously preferable. In any case the important thing is not to be able to attach a specific label to an argument but rather to be able to point out coherently and clearly why the argument in question suffers from confusion or is otherwise unsound.

Although the six fallacies under ambiguity are sometimes considered of minor importance, I have found them extremely valuable in impressing students with the importance of language and in providing them with opportunities to learn to explain themselves clearly and precisely. Students' first efforts at this, as every teacher can testify, are surprisingly weak and disappointing. They quickly recognize this and are eager to learn.

Just as confusing language deceives in the case of the fallacies of ambiguity, so their misleading resemblance to valid argument forms (achieving this deception by misrepresenting the facts disclosed in the argument) deceives in the case of the fallacies of presumption. The argument

> Exercise is good; Jones should do more of it; it will be good for him.

looks deceptively like the classic argument

> All men are mortal; Socrates is a man; therefore Socrates is mortal.

However, they differ in that "Exercise is good" is an unqualified generalization that may not apply to Jones, who may suffer from a heart condition and specifically been told by his doctor not to exercise. What is most characteristic of the fallacies of presumption is that facts relevant to the argument have not been represented correctly in the premises. This inappropriate treatment of the facts can take three forms: we may (1) overlook, (2) evade, or (3) distort significant facts.

Dividing the ten main fallacies of presumption into these three subclasses has proven effective in my teaching. Under *overlooking the facts*, I discuss sweeping generalization, hasty generalization, and bifurcation. Under *evading the facts*, I deal with begging the question, question-begging epithets, complex question, and special pleading. Under *distorting the facts*, I take up false analogy, false cause, and irrelevant thesis.

Just as a greater awareness of language is the focus of my discussion of the fallacies of ambiguity, my discussion of the fallacies of presumption focuses on developing a great awareness and appreciation of precision in thought. This appreciation is the main benefit of working with the fallacies of presumption, and it arises from the opportunities these fallacies offer to sharpen the student's ability to recognize differences and to make finer distinctions. This is obvious when dealing with such a fallacy as false analogy, but the other fallacies in this category provide similar opportunities. Thus, for example, in sweeping generalization, I teach students to isolate and distinguish as clearly as possible the particular rule or generalization used in the argument ("Exercise is good"). Having done so, I point out how much easier it now is to see whether this rule or generalization, clearly set off from the rest, applies to the case at hand or why it does not. After practicing on fifteen to twenty examples, students become adept at this, with obvious satisfaction with their newfound ability.

What tends to confuse in the fallacies of relevance is the emotional storm the speaker raises, for as those who use these devices know, when feelings run high almost anything will pass for argument. So just as the focus of my discussion of the fallacies of ambiguity is language and the focus of my discussion of the fallacies of presumption is thought, the focus of my discussion of the fallacy of relevance is emotion and the havoc it often plays with our thinking. I have found it interesting how the six fallacies traditionally discussed in this connection cover such a wide spectrum of emotion: our susceptibility to prejudice in the fallacy of personal attack, flattery and envy in the fallacy of mob appeal, sympathy in appeal to pity, vanity in appeal to authority, pride in appeal to ignorance, and intimidation in appeal to fear. What a student gains after dealing in great detail with the various examples offered here, many of which I have used in my own classes over the years, is not only a greater ability to distinguish between appeals directed to our reason from those directed to our emotions, which is the primary goal here, but also a deeper knowledge of our emotional nature.

It is a pleasure to be able to acknowledge here the help received by various people while preparing this book for publication. I am especially grateful to Charlie Busch, Jr., a dear friend and former teaching assistant for running down references and helping with various answers; to Patty Holland, a former student and a dear friend too, for spending a good part of the summer of 1978 typing one draft of the manuscript after another; and to Judy Brown, editor in Prentice-Hall's Redwood City office, whom I hardly know how to thank except perhaps to say that it was a great joy working with her. Now that the work is finished I shall miss her calls.

Analyzing
Informal Fallacies

PART ONE

Fallacies
of Ambiguity

Amphiboly

Amphiboly is the first of the six fallacies of ambiguity. These fallacies arise from the confusing way we sometimes express ourselves. In the case of amphiboly this comes about from the careless way we put our words together in sentences, thus giving rise to meanings we had not intended to convey.

*1.

"I'm not going to send you to prison for attempted robbery.
I'm going to give you a second chance!"

2. Student paper:

It all came back to him so plainly: a picture of the loveliest girl he'd ever seen hanging in the navy locker.

*Answers to examples marked with an asterisk are at the end of each part.

3. Announcement:

It would be a great help toward keeping the churchyard in good order if others would follow the example of those who clip the grass on their own graves.

*4. Advertisement:

Mrs. Manning's are the finest pork and beans you ever ate. So when you order pork and beans, be sure Mrs. Manning is on the can.

5. Title of a book by Maurice Zolotow:

No People Like Show People
New York: Random House, 1951.

6. Advertisement:

Wanted Smart Young Man For Butcher. Able to Cut, Skewer, and Serve a Customer.

*7. Sign on store window:

Bathing Suits 40 Percent Off

8. Headline:

Nude Patrol OK'd for Muir Beach

9. Advertisement:

FOR SALE: Antique desk suitable for lady with curved legs and large drawers.

*10. Vital statistic:

Cause of Death: Don't know. Died without the aid of a physician.

11. Vital statistic:

Had never been fatally ill before.

12. **Vital statistic:**

Went to bed feeling well but woke up dead.

13. **Report of social worker:**

Woman still owes $45 for a funeral she had recently.

*14. **News item:**

Alderman Simms said the council ought to be given the whole truth that there was still sufficient coal in the city to last five weeks if nobody used it.

15. **News item:**

Young Lusk, officers said, was found in the trunk of his car in the parking lot of the Disneyland Hotel. He said he had been there almost thirteen hours. The youth said he had tried unsuccessfully to open the trunk lid and finally went to sleep. "I was pretty scared," he told officers. He said he never got a look at his kidnappers, "but I think I would recognize him if I saw him again."

*16. **Editorial:**

Unfortunately the prime minister had left before the debate began. Otherwise he would have heard some caustic comments on his absence.

17. **Announcement:**

Dr. W. T. Jones read an interesting paper entitled "Idiots from Birth." There were over two hundred present.

18. **Announcement:**

Another performance of the pantomime is to be given in the Temple Auditorium. This will give those who missed it another chance of doing so.

*19. **Advertisement:**

A Superb Restaurant. Fine Foods Expertly Served By Waitresses in Appetizing Forms.

*20. Shakespeare's *Henry VI, Part II,* scene 4:

Spirit: The Duke yet lives that Henry shall depose.

21. Record featuring Peter Best, a former member of the Beatles:

Best of the Beatles

*22. A wife, who has been looking at some travel folders with her husband, turns and says,

Darling, if one of us dies, I think I shall go and live in Paris.

23. A husband, away on a business trip, writes the following card to his wife:

Having a wonderful time. Wish you were her.

*24. Announcement:

The accused will be given a fair trial before he is hanged.

25. Advertisement:

Raincoats At Less Than Cost Price. Last Three Days.

26. From Kant:

In this manner, then, results a harmony like that which a certain satirical poem depicts as existing between a married couple bent on going to ruin, "O, marvellous harmony, what he wishes, she wishes also;" or like what is said of the pledge of Francis I to the Emperor Charles V, "What my brother Charles wishes that I wish also" (viz. Milan).

Immanuel Kant, *Critique of Practical Reason and Other Works on the Theory of Ethics,* trans. T. K. Abbott (London: Longmans, Green, 1927), p. 116.

Accent

Unintended meanings can arise not only from faulty sentence structure, as in the case of amphiboly, but also from confusion as to emphasis, as in the case of *accent*. The fallacy of accent results when (1) a statement is spoken in a tone of voice not intended for it; (2) certain words in it are wrongly accented or stressed; or (3) certain words (or even whole sentences and paragraphs) are taken out of context and thus given an emphasis (and therefore a meaning) they were not meant to have.

The following news report regarding the Nixon tapes indicates the important dimension of this fallacy:

> As the transcripts of the Watergate tapes tell it, when John Dean warned Richard Nixon against getting involved in a cover-up, the President answered: "No—it is wrong, that's for sure." But just what inflection was in Nixon's voice when he made this remark, or the many other similarly intriguing but ambiguous bits of dialogue quoted in the transcripts? The public will have to wait to find out. Last week the Supreme Court refused to turn over the 22 hours of Nixon tapes that were played at the Watergate cover-up trial to Warner Communications, the broadcasting networks, public television, and a new directors association, which had been seeking them since 1974.
>
> "Tape Tie-Up. Nixon Wins a Delay." *Time,* May 1, 1978, p. 72. Reprinted by permission from *Time, The Weekly Newsmagazine;* Copyright Time Inc. 1978.

Exercises 27 to 53 are examples of the three forms this fallacy assumes.

27. **Promotion for an upcoming article about women:**

"What," we said with narrowed eyes, "is a *real* woman?"

*28. **Review:**

Only Hollywood could produce a film like this.

*29. **Saying:**

You never looked better!

30. **Remark:**

I wish you all the good fortune you deserve!

31. **Member of audience after sitting through a five-hour performance of Wagner's opera *Parsifal:***

"I can't believe I heard the whole thing."

32.   The ending of the mass in some versions:

Priest: The mass is ended.
Congregation: Thanks be to God.

33.   Reply:

I don't know.

*34.   Question:

Are you going home this weekend?

35.   Question:

What are you doing this weekend? The usual?

*36.   Remark:

You're sober now.

37.   Newspaper headline:

"Perfect Officer" Pressured Out of the Military, He Charges

38.   Brother replying to his sister after being thanked for helping her sew a button for her:

Don't mention it.

39.   Remark:

She didn't try to commit suicide today.

40.   Federal regulation:

Warning: Under Title 18 U.S. Code: It is a Federal Offense to Assualt a Postal Employee while on Duty.

41.   Maxim:

We should not speak ill of our friends.

42. Maxim:

Be courteous to strangers.

43. Overheard remark:

Do stop making that same noise!

*44. Sitcom dialogue:

Those people seem to be walking very carefully, so I needn't slow down for them. After all I was only told to watch out for careless pedestrians.

*45. From an essay:

They will be married Sunday. Then they will spend a few weeks in a cottage by the sea, and by the time the honeymoon is over the groom hopes to be in the army.

46. Inscription on a tombstone:

Sacred to the memory of. . . .
After living with her husband for fifty-five years,
she departed in hope of a better life.

47. School signs:

Slow School

Slow School

*48. Advertisement:

Free Hand Drawing on Genuine Canvas

*49. The czar's reply to a prisoner's plea for a pardon:

Pardon Impossible To Be Executed

*50. Politician's remark:

This is a fine country to live in.

*51. Shakespeare's *Macbeth,* act 1, scene 7:

Macbeth: If we should fail?
Lady Macbeth: We fail.

*52. The teacher, having told her class that "communism is the best type of government if you care nothing for your liberty or your material welfare," is reported by Johnny to his parents at dinner to have said:

Do you know what Ms. Jones said today? She said communism is the best form of government!

53. A bumper sticker put out by the Republicans in the 1972 campaign:

Will Rogers Never Met George McGovern

Hypostatization

To *personify* is to attribute to things and animals qualities that, strictly speaking, are applicable only to human beings. It is to say such things as (to quote a recent headline) "Hurricane Inez Aims Smashing Blow at Texas Coast," which implies a hurricane can have forethought, take sight ("aim") at some goal, and so forth. To *hypostatize* is to treat ideas or concepts in this manner. To use personification in this sense is to make it stand for both *animism* (ascribing such qualities to things) and *anthropomorphism* (ascribing such qualities to non–human beings). Exercises 54 to 74 are examples of this fallacy.

*54. Saying:

The Law Commands

55. Remark:

Science has not produced the general happiness that people expected, and now it has fallen under the sway of greed and power.

56. Editorial:

Today big and complicated government has a hand in everybody's business and another in every person's pocket.

57. Saying:

A corporation has no soul.

*58. Remark:

The world will no longer laugh.

*59. Naval order given by Lord Nelson (1758–1805) before the Battle of Trafalgar (October 1805), which spelled the end of Napoleon's sea power:

England expects every man to do his duty!

60. Headline:

Industry Struggles in the Pincers of Inflation

61. Editorial:

Because the dumping ground was so far out to sea, the city believed it would never hear from its sludge again, but the city was wrong. The mass of goo slowly grew and sometime around 1970 it began to move, oozing back to haunt New York and the beaches of Long Island.

62. Editorial:

What Americans need is patience. A water well goes dry. Nature replenishes it by sending a good rain. Nature shall replenish the great oil reservoirs. Scientists say it takes a few million years. All we need is patience.

63. Shakespeare's *Julius Caesar,* act 5, scene 5:

Nature might stand up
And say to all the world. "This was a man!"

64. **From "Let Not Women E'er Complain" (1794), stanza 1, by Robert Burns:**

Let not women e'er complain of
Inconstancy in love!
Let not women e'er complain
Fickle man is apt to rove!
Look abroad thro' Nature's range,
Nature's mighty law is change!
Ladies, would it not be strange
Man should then a monster prove?

The Complete Poetical Works of Burns (Boston: Houghton Mifflin, 1897), p. 274.

65. **The following couplet, written (c. 1730) by Alexander Pope, was intended as an epitaph for Newton in Westminster Abbey:**

Nature, and Nature's laws lay hid in night.
God said, "Let Newton be!" and all was light.

The Poems of Alexander Pope, ed. John Butt (London: Methuen, 1963), p. 808.

66. **Eddie Cantor, who had a running feud throughout his career with Georgie Jessel, requested the following inscription be placed on his gravestone when he dies:**

Here in Nature's arms I nestle
Free at last from Georgie Jessel

67. **A tombstone with a rather different sort of inscription:**

Death is a debt
To Nature due
I have paid mine
And so must you.

68. **Circular:**

Nature is very deep and mysterious in its operations and contains many unsounded depths.

*69. From *Social Policy* (March/April 1977), Vol. 7, No. 5:

"THE SYSTEM IS THE SOLUTION"
—AT&T

"THE SYSTEM IS THE PROBLEM"
—THE PROGRESSIVE

If you are one of the growing number of Americans who realize that AT&T's slogan means, "What's good for big business is good for America," then...

...welcome to The Progressive, the monthly magazine that knows it's long past time to make fundamental changes. More and more of us see that

—The System squanders our nation's wealth.

—The System rapes our natural and human environments.

—The System pours hundreds of billions of dollars down a rathole called "national security."

—The System puts profit ahead of people.

The System works, all right—it works for AT&T and Lockheed, for IBM and Exxon—but it doesn't work for us, the American people.

Courtesy The Progressive, Madison, Wisconsin.

*70. Bumper sticker:

Freedom is not only worth fighting for; it is also worth dying for.

*71. Poster:

Love Nature.

72. Letter to the editor:

The secret army drug experiments were shocking. It seems that our government "of the people, by the people, and for the people" is doing it *to* the people. Our bureaucratic system is one big cover-up, leaving the American people in the dark. It's like the old dog's body that grew too big for his legs. He couldn't scratch all the fleas.

73. From British philosopher Ryle:

A foreigner visiting Oxford or Cambridge for the first time is shown a number of colleges, libraries, playing fields, museums, scientific departments and administrative offices. He then asks, "But where is the University? I have seen where the members of the Colleges live, where the Registrar works, where the scientists experiment and the rest. But I have not yet seen the University in which reside and work the members of your University."

Gilbert Ryle, *The Concept of Mind* (New York: Barnes & Noble, 1949), p. 16.

74. An account of another entity ("property") whose nature and status is prey to the same confusion:

The most rigid defenders of the momentary legal definition of "property" apparently think "property" refers to something as substantive as atom and mass. But every good lawyer and every good economist knows that "property" is not a *thing* but merely a verbal announcement that certain traditional powers and privileges of some members of society will be vigorously defended against attack by others. Operationally, the word "property" symbolizes a threat of action; it is a verb-like entity, but (being a noun) the word biases our thought toward the substantives we call *things*. But the permanence enjoyed by property is not the permanence of an atom, but that of a promise (a most unsubstantial thing). Even after we become aware of the misdirection of attention enforced by the noun "property," we may still passively acquiesce to the inaccuracy of its continued use because a degree of social stability is needed to get the day-to-day work accomplished. But when it becomes painfully clear that the continued unthinking use of the word "property" is leading to consequences that are obviously unjust and socially counterproductive, then we must stop short and ask ourselves how we want to re-define the rights of property.

Garrett Hardin, foreword to Christopher D. Stone, *Should Trees Have Standing? Toward Legal Rights for Natural Objects* (New York: Avon Books, 1975), pp. 6–7.

Equivocation

The fallacy of *equivocation* arises when a key term in an argument is allowed to shift its meaning in the course of the argument. The result is that the argument is no longer concerned with the same thing, dealing with one thing in the premise and another in the conclusion.

*75.

76. **From a logic text:**

Everything that runs has feet. The river runs. Therefore the river has feet.

*77. Sitcom dialogue:

John: All men are created equal.
Mary: Then why isn't my opinion worth anything in this house?
John: Because you're not a man, obviously.

78. **Sitcom dialogue:**

Helen: Was he mad because you spilled your coffee on him?
Jack: Yes, he was.
Helen: Then you should have had him locked up like any other madman.

79. **From Shakespeare's *Hamlet*, act 2, scene 2:**

Polonius: What do you read, my lord?
Hamlet: Words, words, words.
Polonius: What is the matter, my lord?
Hamlet: Between who?
Polonius: I mean, the matter that you read, my lord?
Hamlet: Slanders, sir. . . .

80. **From Shakespeare's *Hamlet*, act 5, scene 1:**

Hamlet: Whose grave's this sirrah?
Clown: Mine, Sir. . . .
Hamlet: I think it be thine indeed; for thou liest in't.
Clown: You lie out on't, sir and therefore 'tis not yours;
           for my part, I do not lie in't and yet it is mine.

81. **From Shakespeare's *Twelfth Night*, act 2, scene 3:**

Sir Toby Belch: To be up after midnight, and to go to bed then
                  is early; so that, to go to bed after midnight
                  is to go to bed betimes.

*82. **Overheard remark:**

Birth control is race suicide, for when no children are born the
human race must die out.

83. **Student test:**

Anthropology is the science of man embracing woman.

*84. **Editorial:**

I do not believe in the possibility of eliminating the desire to
fight from humankind, because an organism without fight is
dead or moribund. Life consists of tensions; there must be a
balance of opposite polarities to make a personality, a nation, a
world, or a cosmic system.

*85. From a logic text:

Some birds are domesticated. My parrot is domesticated. My parrot is therefore some bird!

86. Sitcom dialogue:

Diamonds are seldom found in this country, so be careful not to mislay your engagement ring.

87. Sitcom dialogue:

Good steaks are rare these days, so don't order yours well done.

88. During the course of a heated dispute:

Your argument is sound, nothing but sound.

89. American Cancer Society advertisement:

Courtesy American Cancer Society

90. Advertisement:

Merrill Lynch is bullish on America
Courtesy Merrill Lynch Pierce Fenner & Smith, Inc.

91. Advertisement:

Mexico won't leave you cold. We guarantee it.
Courtesy Mexican National Tourist Council.

92. Billboard:

Milk does something for every body.
Courtesy California Milk Advisory Board.

93. Advertisement:

Fly me. I'm Joan.
Courtesy National Airlines, Inc.

*94. From a logic text:

There are laws of nature. Law implies a law-giver. Therefore there must be a cosmic law-giver.

95. Letter to the editor:

One question that many non-Christians often ask me is, "How can there be a law of God that is universal?" This skepticism about universal laws is actually rather silly once we think about it. There are many universal laws that we all accept as a part of natural science. Doesn't water always run downhill? Don't negative electric forces always attract? Why, then, is it so hard to believe that one must accept Christ in order to be redeemed? This is a very simple universal law and sums up the whole of Christian ideology in a nutshell.

*96. From a logic text:

I have the right to publish my opinions concerning the present administration. What is right for me to do I ought to do. Hence I ought to publish them.

97. Sitcom dialogue:

> *Jane:* That old copper kettle isn't worth anything. You can't even boil water in it.
>
> *Mary:* It is worth something. It's an antique.

98. From a debate in Congress:

Anyone who is considered old enough to go into the army and fight for his country is a mature person, and anyone old enough to vote is a mature person too. Hence, anyone old enough to fight is old enough to vote.

*99. Letter:

In our democracy all men are equal. The Declaration of Independence states this clearly and unequivocally. But we tend to forget this great truth. Our society accepts the principle of competition. And competition implies that some men are better than others. But this implication is false. The private is just as good as the general; the file clerk is just as good as the corporation executive. The scholar is no better than the dunce; the philosopher is no better than the fool. We are all born equal.

*100. Editorial:

We are told that to discriminate against women in employment is wrong. Yet we must discriminate. We discriminate when we say that a woman couldn't be a bridge builder or a longshoreman. They simply lack the necessary strength for these tasks. We discriminate when we require college degrees or special training. Hence, it makes no sense to say that discrimination against women is wrong.

*101. Editorial:

It is certainly true that human beings are of the animal kingdom, and it is natural for an animal to exhibit characteristic patterns of behavior. However, human society restricts people's animalistic behavior and so human society is, to that extent, unnatural.

102. **Editorial:**

I am puzzled by the protest groups that gather in front of prisons when an execution is scheduled. The murderer, who has committed a heinous crime, has been granted all due process of law and is given every opportunity to defend himself—usually with the best available legal minds and often at taxpayers' expense. Yet these same protestors generally favor the execution of millions of innocent babies by abortion.

The Spotlight. Published by Liberty Lobby. July 16, 1979, p. 27.

103. **Student Poster:**

MIDTERMS?
GUESS WHAT GOD SAID HE'D DO FOR YOU?

He's promised to increase your knowledge ten-fold and bring all things to your remembrance!
Find out how to apply it at:
REJOYCE IN JESUS BIBLE STUDY
Tuesdays 12:30
SAC 203.

104. **British empiricist philosopher John Locke's argument that there is no such thing as innate knowledge:**

For first, it is evident that all children and idiots have not the least apprehension or thought of them; and the want of that is enough to destroy that universal assent which must needs be the necessary concomitant of all innate truths: it seeming to me near a contradiction to say that there are truths imprinted on the soul which it perceives or understands not; imprinting, if it signify anything, being nothing else but the making certain truths to be perceived. For to imprint anything on the mind without the mind's perceiving it, seems to me hardly intelligible. If therefore children and idiots have souls, have minds, with those impressions upon them, they must unavoidably perceive them, and necessarily know and assent to these truths; which since they do not, it is evident that there are no such impressions....He therefore that talks of innate notions in the understanding, cannot (if he intend thereby any distinct sort of truths) mean such truths to be in the understanding as it never perceived, and is yet wholly ignorant of. For if these words (to be in the understanding) have any propriety, they signify to be understood, so that to be in the understanding and not to be understood, to be in the mind and never to be perceived, is all one as to say anything is and is not in the mind or understanding.

John Locke, "Essay on Human Understanding," in *Modern Classical Philosophers,* ed. Benjamin Rand (Boston: Houghton Mifflin, 1952), pp. 219–20.

105. Twentieth-century philosopher Wittgenstein pointing out one of the typical ways we are misled by language, in this case the language of science:

We have been told by popular scientists that the floor on which we stand is not solid, as it appears to common sense, as it has been discovered that the wood consists of particles filling space so thinly that it can almost be called empty. This is liable to perplex us, for in a way of course we know that the floor is solid, or that, if it isn't solid, this may be due to the wood being rotten but not to its being composed of electrons. To say, on this latter ground, that the floor is not solid is to misuse language. For even if the particles were as big as grains of sand, and as close together as these are in a sandheap, the floor would not be solid if it were composed of grains. Our perplexity was based on a misunderstanding; the picture of the thinly filled space had been wrongly *applied.* For this picture of the structure of matter was meant to explain the very phenomenon of solidity.

Ludwig Wittgenstein, *Blue and Brown Books* (New York: Harper & Row, Pub., 1965), p. 45.

Division

The fallacy of *division* arises when we try to apply what is true of the whole to what we therefore believe must be true of each part or when we try to apply what is true of a group to what we therefore believe must be true of each of its members. These are fallacies because we cannot validly extend properties from wholes to parts or from groups to members, for a group or a whole is something functional. To make such an inference is to neglect the element of structure that makes a whole other than the sum of its parts.

Exercises 106 to 121 are examples of division.

106. Advertisement:

> **See America's favorite**
> **upright vacuum cleaner at**
> **your Hoover dealer. Like**
> **millions of women, you'll**
> **like the basic Hoover and**
> **its honest value.**

Courtesy of The Hoover Company, North Canton, Ohio.

*107. Announcement:

All donors have contributed $1,000.

*108. Old riddle:

Question: Why do white sheep eat more than black ones?
Answer: Because there are more of them!

*109. Interviewer:

I am interested in hiring only the kind of man who will be efficient. Jones simply does not deserve to be classed as such a man. He worked for the civil service five years, and everybody knows the civil service is a notoriously inefficient organization.

110. Letter to the editor:

America is the wealthiest nation in the history of the world; therefore it is absurd to say that poverty is a problem for Americans.

*111. From a logic text:

Carrier pigeons are practically extinct. This bird is a carrier pigeon and is therefore practically extinct.

112. **Overheard remark:**

According to the insurance statistics based upon mortality tables, my life expectancy is twenty years. Therefore I shall die twenty years from now.

*113. **Overheard remark:**

Since every third child in New York is a Catholic, Protestant families there should have no more than two children.

*114. **From Robert Gillette in the *Los Angeles Times*, "The Ban on Saccharin: How? Why?" March 20, 1977, p. 1:**

Elusive as the logic of this law may seem to the general public, medical researchers say that the animal experiments upon which such regulatory decisions are based do have a sensible rationale—and that they do have a direct bearing on risk to humans.

It is true, researchers acknowledge, that the rats at the center of this furor consumed each day the saccharin equivalent of roughly 1,000 cans of low-calorie soft drink. But they consider it misleading—if not dishonest—to suggest, as the calorie council has, that this disparity makes the research irrelevant to humans.

People who drink only one can a day of a saccharin soft drink run a comparably lower risk of cancer. But in a population of 213 million people, the collective hazard could still add up to several thousand cases of bladder cancer.

By Robert Gillette. Copyright, 1977, *Los Angeles Times*. Reprinted by permission.

*115. **From a logic text:**

All the trees in the park make a thick shade; this tree is one of them and therefore makes a thick shade.

116. **Interviewer:**

All colleges of business administration place increasing emphasis on quantitative methods of business decisions, and so you'll find them all introducing students to such things as game theory and linear programming.

117. **News item:**

The American economy has made great progress since 1960 in the quantity of goods and services produced. The shoe industry, being a part of the American economy, has done so too then.

118. **From a logic text:**

Salt is not poisonous, so neither of the elements of which it is composed—sodium and chlorine—is poisonous either.

119. **Student remark:**

The instructor in our physics class told us that all things are made of very small particles or waves. Headaches, dreams, doubts, thoughts, and the like, must also therefore be made up of small particles and waves.

120. **Gambler:**

I have flipped a coin seven times and got seven heads. My next toss should be heads too.

*121. **News item:**

It is predicted that the cost-of-living index will rise again next month. Consequently you can expect to pay more for butter and eggs next month.

Composition

The fallacy of *composition* is the reverse of division: attempting to apply to the whole or group what is true only of the parts or members.
Exercises 122 to 145 are examples of the fallacy in this form.

122. **Textbook example:**

One grain of sand does not make a heap; a second grain of sand added to the first does not make a heap; indeed each and every grain of sand, when added to the others, does not make a heap which was not a heap before. Therefore, all grains of sand taken together would not make a heap.

*123.

Courtesy North American Philips Corp.

124. **A variation of No. 122:**

Atoms are exceedingly small. In fact, they cannot be seen, for they do not provide sufficient stimulus for the optic nerve. If this paper therefore is composed of invisible substances, must it not then also be invisible?

125. **An absurd example:**

There is not a single branch of education that, when considered by itself, can with truth be said to be indispensable. Can we therefore resist the conclusion that we may dispense with education altogether?

*126. **Editorial:**

It is not going to help the energy crisis to have people ride buses instead of cars. Buses use more gas than cars.

127. **Textbook example:**

This story must be well written, for every sentence in it is well written.

128. **Variation of the above:**

I don't see why you criticize the novel as implausible. There isn't a single incident in it that couldn't have happened.

*129. **Overheard:**

You need have no fear of the patient's life; he has received many injuries, but none of them is serious when considered individually.

*130. **One student to another:**

Jones can run the half mile in two minutes, therefore he can run the mile in four minutes.

131. **From Aristotle:**

Should we not assume that just as the eye, the hand, the foot, and in general each part of the body clearly has its own proper function, so man too has some function over and above the function of his parts?

Aristotle, *Nicomachean Ethics,* trans. Martin Ostwald (Indianapolis, Ind.: Bobbs-Merrill, 1962), p. 16.

132. From author Jose Silva:

Can the universe think about itself? We know that at least one part of it can: we ourselves. Is it not reasonable to conclude the whole can?

The Silva Mind Control Method (New York: Pocket Books, 1978), p. 116.

133. Textbook example:

Since all club members have paid their bills, the club must be out of debt.

134. Another:

No one of the crowd would by himself stoop to so mean an act; therefore I am quite sure that as a body they would not do it.

*135. Overheard:

No one of this committee is especially outstanding in ability. It is impossible for the committee, therefore, to bring in an able report.

136. Student comment:

We will have a championship team next year as well because the coach says he is in the process of gathering together spectacular players.

*137. Adapted from Montaigne:

The pawnbroker thrives on the irregularities of youth; the merchant on a scarcity of goods; the architect and contractor on the destruction of buildings; lawyers and judges on disputes and illegalities; the military on war; physicians on sickness and morticians on death. If, then, we have more profligacy, destruction, lawlessness, war, disease, and death, we shall have unparalleled prosperity.

Montaigne, "That One Man's Profit is Another's Loss," in *The Complete Works of Montaigne,* trans. Donald M. Frame (Stanford, Calif.: Stanford University Press, 1957), pp. 76–77.

138.   Overheard:

You say the United States has the highest living standard of any nation in the world? I can disprove that statement by pointing to the sharecroppers in the South. Is that what you mean by a high standard of living?

*139.   Letter to the editor:

If we lower income tax, everyone will benefit and the country will be better off.

*140.   Student comment:

Any student in college would stand higher in class if he or she received higher marks; hence if all marks were raised 10 percent, everyone would stand nearer the head of the class.

141.   Overheard:

Money is the source of happiness and comfort. If everybody had more money, the whole world would be a happier and more comfortable place.

142.   Politician:

Most hung juries result from the obdurate attitude of one or two persons. Thus I propose to eliminate the juridical problem of hung juries by composing them of not more than ten persons.

*143.   Textbook example:

Some day man will disappear from earth, for we know that every man is mortal.

144. From Albright:

Every human group is born, grows, declines, and dies, as it must if it is an aggregation of individual living beings.

William F. Albright, *From the Stone Age to Christianity: Monotheism and the Historical Process* (Baltimore: Johns Hopkins Press, 1940), p. 80.

145. From Mill:

No reason can be given why the general happiness is desirable, except that each person, so far as he believes it to be attainable, desires his own happiness. This, however, being a fact, we have not only all the proof which the case admits of, but all which it is possible to require, that happiness is a good, that each person's happiness is a good, that each person's happiness is a good to that person, and the general happiness, therefore, a good to the aggregate of persons.

John Stuart Mill, *Utilitarianism* (1863), ed. Oskar Piest (Indianapolis, Ind.: Bobbs-Merrill, 1957), pp. 44–45.

ANSWERS TO EXERCISES
PART ONE

Amphiboly

1. The judge will give him a second chance to do what? Using a term (often a pronoun) whose antecedent is somewhat ambiguous is one of the main causes of this fallacy.

4. Failure to adjust our words to changing contexts is another source of the fallacy. Sometimes this can be avoided by the addition of another word, here by writing "be sure Mrs. Manning's *picture* is on the can."

7. Another source of the fallacy is trying to be too brief, as in this example. Off what? The suit or the price?

10. Emotional stress can play havoc with our hold on grammar, as in this example. Does one need a physician to help one die?

14. Just plain ordinary clumsiness can give rise to the fallacy too. In this case if nobody used the coal, it would last forever.

16. The fallacy can also arise from the incongruous juxtaposition of two sentences. Here if the prime minister had not left, there would have been no reason to complain.

19. The fallacy is sometimes exploited for its humor (in which case, of course, it is not a fallacy, for to feign to commit a fallacy is not to commit one).

20. Sometimes the fallacy is exploited for its very ambiguity as here. Who is going to depose whom? Will changing the word "that" to "who" or "whom" tell us?

22. The fallacy teaches us we should be careful to construct our sentences in such a way that they do not give rise to a meaning we do not intend to convey. The psychological dimension of the fallacy shows that such constructions sometimes also betray an intention or a desire we have failed to repress successfully or kept from surfacing. Allowing our thoughts to travel heedlessly on the lines laid down by common expressions ("if one of us dies, the other..." becoming "if one of us dies, I...") allows our unconscious to interpose itself, taking advantage of our momentary distraction.

24. To avoid the implication that they intend to hang him whether he is guilty or not, the remark would have to be restated as follows: The accused will be given a fair trial and, if found guilty, he will be hanged.

Accent

28. How does the tone of voice used in saying this change the meaning of the remark (from "only Hollywood could produce a bad film like this" to "only Hollywood could produce a good film like this")?

29. Does that mean you were never more beautiful than now? Or does it mean you always looked that way (that is, bad)? It will depend on the tone of voice—whether serious or ironic—in which it is spoken.

34. Said with excitement or in a dreary tone of voice?

36. Said optimistically (and now you will stay that way) or pessimistically (but for how long?).

44. Placing too much emphasis on what word here has led to this somewhat shaky inference?

45. How does the word "hopes" draw too much of the wrong kind of attention to itself here? What word might be better?

48. Imagine a period after the word "Free" or a hyphen between "Free" and "Hand" and notice the difference this makes.

49. What difference does the way this message is punctuated (and with it the different places where the stress falls) make to what it conveys?

50. As in the case of amphiboly, the fallacy of accent is sometimes exploited for its humor, as in this example. The politician gets himself out of a difficulty by pleading that the remark he made reflected favorably and not unfavorably on the government.

51. The ambiguity inherent in the fallacy can give rise to real difficulties, as in this example from Shakespeare. Notice the many different ways Lady Macbeth's reply might be uttered:

(1) We fail. *Is that possible?* (Said in a questioning tone)
(2) We fail. *And that will be that!* (With fatalism)
(3) We fail. *Impossible!* (Said sarcastically)

And perhaps in other ways as well.

52. Next day, we assume, matters are clarified in the principal's office. This is the third form in which the fallacy can occur.

Hypostatization

54. The "Law" obviously does not have a voice with which to issue commands. If the term is used merely as a metaphor and the user is aware of this, no harm is committed. Very often, however, this use of language leads us and others to draw fallacious inferences regarding the things so spoken about.

58. How many people have been intimidated by this very fear? At best it is simply a conceptual fiction (a construction), and at worst it can represent only a very few people (for how many people actually know any one of us, or care one way or the other about us?).

59. Why is this so much more intimidating, and therefore more effective, than if he had said, "I expect every man to do his duty"?

69. We engage in this kind of language usage because of the great deal of mileage we derive from it. Thus it is much easier to attack and condemn such abstractions as "the System" or "the Establishment" than to point out what particular laws or rules or practices one feels need changing.

70. As this example shows, we pay a stiff price for the mileage we get in using this kind of verbal magic. Before responding to this kind of rallying call, one should try to reduce the abstraction to something more concrete by asking such specific questions as: Freedom for whom? Freedom from what? Freedom to do what?

71. Whenever we suspect a term is being hypostatized, we should ask for its concrete referent. Does "Love Nature" mean we should love rocks, ants, earthquakes, and hurricanes? The remark may have meaning, but it is not obvious without further elaboration what that is and therefore how sound the contention is.

Equivocation

75. Kind of cheap.

77. Recognizing that the fallacy of equivocation arises from a shift in the meaning of a word in the course of an argument, we can always expose such fallacies by insisting that the meaning of the key terms remain constant. Notice what happens if this is done with the term "men" in this interchange: if John intends that term to stand for "human being," then his final remark is false; and if he intends that term to mean "male" (namely, that only males were created equal), then his first statement is false.

82. The fallacy is especially easy to commit if the key term on which it turns has both a literal and a metaphorical sense, as in this example. What does the author of this argument think "birth control" stands for? What was it meant to stand for?

84. A much more difficult example involving the use of two similar expressions ("desire to fight" and "without fight"), one used literally, the other metaphorically.

85. The fallacy is not confined solely to figures of speech or metaphors. It can result from the equivocal use of almost any word, even a little word like "some," as in this absurd example.

94. An ancient example but still convincing to those who fail to distinguish laws of nature (which are simply observations we have made) from legal or moral laws (which are not observations but commands).

96. "Right" can mean privilege as well as duty.

99. When we declare that all men are equal, do we mean they all possess the same talents and abilities? And if they are not equal in respect to such native abilities, in respect to what do we nevertheless consider them equal?

100. This argument is more difficult. Are all "discriminations" alike, or are some somewhat more arbitrary (and therefore not justifiable) than others?

101. How does the meaning packed into the term "animalistic" in this argument differ from the meaning conveyed by the term "animal"? Are the meanings equivalent?

Division

107. Confusion between the collective and distributive senses of key terms ("all" in this example) is what frequently lies behind the tendency to commit this fallacy. If all together contributed $1,000 and you incorrectly inferred from this statement that each one did so, you would have fallen victim to this fallacy.

108. The riddle treats collectively in the answer what was referred to distributively in the question: the question asked why each and every white sheep ate more than each and every black one; in the answer we are told this is so because together there are more of them.

109. Just because Jones worked for an inefficient organization is no proof that he was or is inefficient. He may have been the exception.

111. This may be a case of equivocation as well. "Extinct" applied to a group means they are getting fewer; but applied to an individual, it means it is dying. This particular bird may be hale and hearty and not dying at all.

113. The absurdity results from going from what is true of a "a third of the child population of New York" to "the third child born." A fourth of the world's population may be Chinese, but this does not mean that every fourth child born is therefore Chinese.

114. A beautiful example of the fallacy. The fact that there are 213 million people in the country and that therefore a large number of them drink diet drinks does not affect the amount each one drinks. Each person still drinks the amount he or she does, one remarkably less than the equivalent fed the rats in question.

115. Confusion again, perhaps, between the collective and distributive meanings of the term "all." The tree in question may have very few leaves and offer little shade. This and the remaining examples are designed to illustrate the fallacy involving wholes and parts.

121. An example of the misuse of averages. The cost-of-living index is something arrived at by averaging out a great number of different items. This being so, we cannot tell in advance what the case will be with any specific one. Butter and egg prices may also rise, but then they may stay the same or even go down.

Composition

123. Will twenty-two blades be better still? Or is eighteen too many already? And is it meaningful to try to compare two such different types of shavers?

126. A much more difficult case, involving confusion over the distributive and collective senses of the terms buses and cars. Each bus, of course, uses more gas than each car; but since there are a great many more cars than buses, it will indeed be a help if more people ride buses.

129. Happening to him all at once may so sap his strength that he may not survive the ordeal.

130. But will he be able to keep up the pace, or will he tire?

135. Although no one of them may be especially outstanding, by pooling their resources they may together be able to do what no one of them can do individually.

137. If all this took place at more-or-less the same time, in the same place, to the same people, no one would thrive.

139. Everyone may benefit in the sense of having to pay less income tax, but we may be deprived of important services the country may no longer be able to provide.

140. If everyone's grades were raised by the same amount, no one would stand any higher than he or she did before.

143. Although each man is mortal and will die someday, the human race, by replenishing itself each generation, may go on indefinitely. This could be a case of equivocation as well—on the term "man" (in the sense of the individual man and mankind).

TEST ON THE FALLACIES
OF AMBIGUITY

1. Sign:

WE NEVER CLOSE
Open Every Night til Midnight

2. Editorial:

But what are these attackers of patriotism really against? They rise up in fury if anyone suggests they are opposed to patriotism itself; they simply say they dislike "too much" of it. They dislike "extremists" and prefer "moderates." These words and ideas have no sensible content. If patriotism is good, then a lot of it should be extremely good and who would not want something that is extremely good?

3. Inscription on tombstone:

U.S. Economy. Victim of Energy Crisis.
Circa 2000 A.D.

4. Advertisement:

Good health is just a jump away. Jump yourself into good health by following this simple routine each day.

5. Headline:

Lawyers to Offer Poor Free Advice

6. News report:

A child murderer stalks Detroit's frightened suburbs.

7. From Leibniz:

And to judge still better of the minute perceptions which we cannot distinguish in the crowd, I am wont to make use of the example of the roar or noise of the sea which strikes one when on its shore. To understand this noise as it is made, it would be necessary to hear the parts which compose this whole, i.e., the noise of each wave, although each of these little noises would not be noticed if the wave which makes it were alone. For it must be that we are affected a little by the motion of this wave, and that we have some perception of each one of these noises, small as they are; otherwise we would not have that of a hundred thousand waves, since a hundred thousand nothings cannot make something.

G. W. Leibniz, *New Essays Concerning Human Understanding* (1704), trans. Alfred Gideon Langley (Chicago: Open Court Publishing, 1916), p. 48.

8. **Sign:**

We Serve Crabs Here

9. **Advertisement:**

Let the computer teach you about yourself and what the future holds for you.

10. **Overheard:**

The combination that has been formed need not be feared, for all its members are exceedingly weak and inexperienced.

11. **Sign:**

Watch For Falling Rocks

12. **From John Stuart Mill:**

The only proof capable of being given that an object is visible, is that people actually see it. The only proof that a sound is audible, is that people hear it; and so of the other sources of our experience. In like manner, I apprehend, the sole evidence it is possible to produce that anything is desirable, is that people do actually desire it.

John Stuart Mill, *Utilitarianism* (1863), ed., Oskar Piest (Indianapolis, Inc.: Bobbs-Merrill, 1957), p. 44.

13. **Editorial:**

All manufacturers are perfectly free to set their own price on the product they produce, so there can be nothing wrong with all manufacturers getting together to fix the prices of the articles made by all of them.

14. **Sign:**

Fine For Littering

15. **Religious circular:**

How can you doubt that the soul survives the body? Why consider how each spring nature has a new birth. Surely you cannot doubt that what is true of the rest of nature is true of human beings?

16. **Announcement:**

Housewives to Hear Houseplants Talk

17. **Student essay:**

If each thing has a cause, then the whole world and all human history has a cause.

18. **Letter to the editor:**

We are told that prostitution is a growing national problem, but that isn't the half of it! At least half the men and women in this country today are prostitutes. They sell their bodies or their minds on jobs that are personally meaningless and socially destructive.

19. **Airline flight attendant:**

Please fasten your seat belt as we will be landing shortly.

20. **From *The Notebooks of Lazarous Long* by Robert A. Heinlein:**

There are hidden contradictions in the minds of people who "love Nature" while deploring the "artificialities" with which "Man has spoiled 'Nature.'" The obvious contradiction lies in their choice of words, which imply that Man and his artifacts are nòt part of "Nature"—but beavers and their dams are. Such contradictions go deeper than this prima-facie absurdity. In declaring his love for a beaver dam (erected by beavers for beavers' purposes) and his hatred for dams erected by men (for the purposes of men), the "Naturist" reveals his hatred for his own race—i.e., his own self-hatred.

ANSWERS TO TEST
ON THE FALLACIES
OF AMBIGUITY

1. Amphiboly
2. Composition
3. Hypostatization
4. Equivocation
5. Accent
6. Accent
7. Division
8. Equivocation
9. Hypostatization
10. Composition
11. Accent
12. Equivocation
13. Composition
14. Accent
15. Division
16. Amphiboly
17. Composition
18. Equivocation
19. Amphiboly
20. Equivocation

PART TWO

Fallacies
of Presumption

"I've told you a million times, DON'T EXAGGERATE!"

Sweeping Generalization

The error in the fallacy of *sweeping generalization* lies in assuming that what is true under certain conditions must be true under all conditions. This is the first of the fallacies under the general category of *presumption* (fallacies in which facts relevant to the argument have not been represented correctly in the premises) and the first of the three arranged here under the subclass of *overlooking the facts*.

Some writers treat this fallacy (and the next one, which is its reverse) under the heading of *accident* and *converse accident*; others use the Latin descriptions *argumentum a dicto simpliciter ad dictum secundum quid* (arguing from a statement made simply to a statement made under some special condition) and *argumentum a dicto secundum quid ad dictum simpliciter* (arguing from a statement made under some special condition to a statement made simply).

A frequent simple example of accident is some variant of the following: "This country is a democracy and is dedicated to the proposition that all persons are free and cannot be deprived of their liberties. Therefore we should stop imprisoning criminals and lunatics." Reversing the direction of this argument, we get the following as an illustration of the fallacy of converse accident: "Criminals and lunatics are put in institutions against their will. I could name more. But enough is said to see that our country is not truly dedicated to the proposition that all persons are free and that one of their rights is liberty."

Exercises 146 to 157 are examples of sweeping generalization or accident.

*146. An absurd example:

Hurting another person is bad; therefore dentists commit bad acts.

*147. Another absurd example:

Those who thrust a knife into another person should be punished; surgeons in operating do so; therefore they should be punished.

*148. And still another:

Everyone enjoys a good joke. I am sure, therefore, that Jones won't mind us playing one on him.

149. Cliché:

Rising taxes must come down soon, for all that goes up must come down.

*150. Another:

The president should get rid of his advisors and run the government by himself. After all, too many cooks spoil the broth.

*151. Somewhat more serious:

Narcotics are habit forming. Therefore if you allow your physician to ease your pain with an opiate, you will become a hopeless drug addict.

152. And more difficult:

No man who lives on terms of intimate friendship and confidence with another is justified in killing him. Brutus therefore did wrong in assassinating Caesar.

*153. Still more difficult:

No wealth can be created without labor. Therefore the working classes should get the major portion of that wealth that would not exist without their help.

154. American Secretary of State Cordell Hull, refusing to grant asylum to refugees on the ship the St. Louis, whose forced return to Nazi Germany meant certain death:

I took an oath to protect the flag and obey the laws of my country and you are asking me to break those laws.

155. Socrates explaining to his old friend Crito why, even though he is innocent, he will not try to escape but feels impelled to submit to the penalty imposed on him by the Athenian court:

Compared with your mother and father and all the rest of your ancestors your country is something far more precious, more venerable, more sacred, and held in greater honor both among gods and among all reasonable men. Do you not realize that you are even more bound to respect and placate the anger of your country than your father's anger? That if you cannot persuade your country you must do whatever it orders, and patiently submit to any punishment that it imposes, whether it be flogging or imprisonment? And if it leads you out to war, to be wounded or killed, you must comply, and it is right that you should do so; you must not give way or retreat or abandon your position. Both in war and in the law-courts and everywhere else you must do whatever your city and your country commands, or else persuade it in accordance with universal justice; but violence is a sin even against your parents, and it is a far greater sin against your country.

Plato, "Crito" (Stephanus 50 B-51 B), in *Plato: The Last Days of Socrates*, trans. Hugh Tredennick (Harmondsworth, England: Penguin, 1961), p. 91.

156. News item:

In the anti-Israeli film, "The Palestinian," financed and narrated by the fanatical anti-Semite, Vanessa Redgrave, Yassir Arafat is quoted as saying: "The only solution of the Middle East problem is the liquidation of the state of Israel." Miss Redgrave replies: "Certainly." The "liquidation" of the state of Israel means nothing more nor less than the slaughter of hundreds of thousands of Jewish men, women, and children—a second Holocaust against the Jewish people in one generation. It means, in addition, the destruction of the last best hope for Jewish survival—the Jewish state. This call for genocide was scheduled for a public screening tonight at the Academy of Motion Picture Arts and Sciences' Samuel Goldwyn Theatre in Beverly Hills. It is probably hopeless to attempt to explain to people like Academy president Howard W. Koch, that no one with the slightest spark of human decency would state, as a spokesman for his organization did, that "Anyone who has a film and wants to screen it can rent the theatre for $800."

"The Redgrave Challenge . . ." by Prof. Sol Modell. *Heritage,* Los Angeles, June 16, 1978. Opinion Section D.

157. From author Gloria Emerson:

Mr. Wilson thought I made too much fuss about casualties in Vietnam.

"Now, every man wants to live," he said. "But the trouble is we're all born to die. So what the hell is the difference if we die when we're eighty or if we die when we're twenty-six."

Winners and Losers: Battles, Retreats, Gains, Losses and Ruins from a Long War (New York: Random House, 1976), pp. 261–62.

Hasty Generalization

In *sweeping generalization* a generalization is misapplied. In *hasty generalization* a particular case is misused. It can be misused in the sense that it is not representative of the cases that would warrant the conclusion in question (158 to 166), because of an insufficient number used to arrive at the conclusion in question (167 to 173), because there is no essential connection between it and the generalization it is called in to support (174 to 180), or because important contrary evidence that would tend to cast doubt on the conclusion drawn has not been considered (181 to 186).

*158. An absurd case:

They just don't care about traffic law enforcement in this town. They let ambulances go at any speed they like and let them run red lights, too.

159. Another:

When Bill had to go upstate with the debating team, the professor told him it was perfectly all right for him to miss class. That's why I say the professor doesn't care whether any of us comes to class.

160. And still another:

Judge Wisdom gave a suspended sentence to the boy when she found out this sixteen-year-old youth's father forced him to sell dope. It is clear, then, what I said is true, namely, that Judge Wisdom is soft on dope pushers.

161. Overheard:

The welfare program is totally unnecessary. Why, I know a guy who runs a very lucrative gambling operation and who drives his new Cadillac downtown every week to collect his welfare check.

*162. Overheard:

The newspapers are full of nothing but stories about sex and crime. In last night's paper, for instance, there were five crime stories on the first page.

*163. Overheard:

When you think of what some people have accomplished without a college education, you will agree with me that it will be a waste of time for anyone to spend four years there.

*164. Tire advertisement:

If they're tough enough for Alaska, they're tough enough for you.

From *Car & Driver* Magazine, November 1977.

*165. Remark:

Science is not to be taken seriously. It has not been able to explain the origin of life.

166. Remark:

I spent all morning canvasing for the committee to reelect our representative. Here I was in a typical upper-middle-class neighborhood—prime Republican territory—and yet eleven of the nineteen people I talked to who had made up their minds are going to vote Democratic this year. Well, that does it; no more canvasing for me. Our candidate doesn't have a chance.

*167. Overheard:

The clerks in Mason's Department Store are incompetent. They got two of my orders mixed up during the last Christmas rush season.

*168. Overheard:

Since attending that mass meeting of students yesterday, I have ceased having any confidence in decisions reached in that way.

169. DEAR ABBY:

You said that most male doctors do not get turned on by an attractive female patient. You're all wet! My ex-wife had a thing going with her doctor for a long time before I caught on. I didn't prosecute because I figured the poor guy had enough trouble being emotionally involved with my wife. During the last five years I paid enough doctor bills to put another Cadillac in his garage. Sweetie, you may know a lot about teen-agers, but you had better do a little more research on doctors.

BURNED ONCE

Courtesy Chicago Tribune–New York News Syndicate Inc.

170. News item:

In May of 1968 I was interviewed by two North Vietnamese generals at separate times. Both of them said to me, in almost these words: "After we liberate South Vietnam we're going to liberate Cambodia. And after Cambodia we're going to liberate Laos, and after we liberate Laos we're going to liberate Thailand. And after we liberate Thailand we're going to liberate Malaysia, and then Burma. We're going to liberate all of South-east Asia." They left no doubt in my mind that it was not a question of South Vietnam alone. Some people's favorite game is to refute the "domino theory," but the North Vietnamese themselves never tried to refute it. They believe it.

*171. Superintendent:

American parents are not opposed to school busing programs. We have had such a program in Berkeley for a number of years, and it has worked out quite satisfactorily.

*172. Anthony Burgess in "Is America Falling Apart?" *New York Times Magazine,* November 7, 1971, p. 101.

America has always despised its teachers and, as a consequence, it has been granted the teachers it deserves. The quality of first-grade education that my son received, in a New Jersey town noted for the excellence of its public schools, could not, I suppose, be faulted on the level of dogged conscientiousness. The principal had read all the right pedagogic books, and was ready to quote these in the footnotes to his circular exhortations to parents. The teachers worked rigidly from approved rigidly programed primers, ensuring that school textbook publication remains the big business it is. But there seemed to be no spark; no daring, no madness, no readiness to engage in the individual child's mind as anything other than raw material for statistical reductions. The fear of being unorthodox is rooted in the American teacher's soul: you can be fired for teaching the path of experimental enterprise.

© 1971 by New York Times Company. Reprinted by permission.

*173. DEAR ANN LANDERS:

I'm writing about the letter by "Janus" who was bewildered and hurt by the man she loved because he seemed to be "two people." One day he was the dearest and most considerate of lovers. The next day, for no apparent reason, he was remote and brooding. I would be willing to bet this man is a Pisces, born around Feb. 20 to March 20. These radical personality changes are characteristic of a Piscean and anyone who has had the misfortune to love someone born under this sign of the zodiac must be willing to understand and accept his unpredictable behavior or get out of his life. I can tell you from experience that Pisceans are fascinating but they are not worth the wear and tear on the nervous system. You were right when you told her, "Fight it with your hat. Take it and run."

ARIES IN ANCHORAGE

Courtesy Ann Landers, Field Newspaper Syndicate, the Los Angeles Herald-Examiner.

*174. Edmond Rostand's *Cyrano de Bergerac*, act 1, scene 1:

Cyrano: A great nose indicates a great man.

*175. Nobel laureate Dr. Linus Pauling on President Nixon:

For fifteen years I have studied insanity. I saw the eyes on television, and there is madness, paranoia.

*176. Overheard:

This must be good perfume because it is in a very elegant bottle.

177. **Advertisement:**

178. **Overheard:**

I won't consider a young man for this job—they are unreliable.

*179. **Overheard:**

She is very fond of children and so she will undoubtedly make a fine kindergarten teacher.

*180. **News item:**

Thor Heyerdahl has done it again, crossing the Atlantic in a papyrus raft fashioned after ancient Egyptian tomb carvings. Landing in the Western Hemisphere on the island of Barbados, he was greeted by the Barbados Prime Minister, Errol Barrow, who declared, "This has established Barbados was the first landing place for man in the Western World."

*181. **Remark:**

State-owned industries encourage featherbedding and absenteeism. Thus all state-owned industries should be abolished.

182. **Overheard:**

To fail children who do poorly upsets them and disturbs class morale. The only thing to do is to promote everyone.

*183. **Editorial:**

High tariffs enable our industries to grow strong; they assure high wages to the workers and they increase federal revenues. High tariffs are therefore a benefit to the nation.

*184. **Editorial:**

In the past, America has prospered without government control of labor. There is, therefore, no reason we should start such control now.

185. **Overheard:**

Wine is a natural drink because it is nothing but the juice of grapes.

186. Nobel laureate Dr. Albert Szent-Gyorgyi, discoverer of vitamin C, telling of his experience with some clinicians:

I had hardly announced my result when I was invited by E. Merck and Co. to come to Darmstadt, Germany, to discuss the medical uses of ascorbic acid with leading German clinicians. . . . I found the clinicians very nice but was shocked by their lack of biological insight and lack of interest in the basic problems. They told me that they had no need of my ascorbic acid because ascorbic acid prevented scurvy, and there was no scurvy. The logic was simple, almost irrefutable. Though I could bring up no argument against it, I felt that it was completely wrong.

Executive Health, Vol. 13, No. 9, 1977.

Bifurcation

The term *bifurcation* refers to a fallacy that presumes a certain distinction or classification is exhaustive and exclusive when other alternatives are possible. The fallacy is sometimes referred to as the *either/or fallacy* or the *black or white fallacy*. The fallacy often has its source in confusion over contradictories (dead/alive) and contraries (rich/poor)—a statement consisting of contraries put forward as if it contained contradictories. In some cases of either/or there are indeed no other alternatives ("Jones is either dead or alive"; "You will either pass the test or fail it"); but in the vast majority of cases, a wide spectrum of possibilities is overlooked. Exercises 187 to 223 are examples of bifurcation.

187. Bumper sticker:

America: Love It or Leave It

188. Its reply:

America: Change It or Lose It

*189. TV commercial:

When you're out of Schlitz, you're out of beer.

Courtesy Jos. Schlitz Brewing Company.

190. Advertisement:

Black & White. Wherever life is lived with taste and style, you'll find the brilliant taste of Black & White.

Courtesy Heublein Inc.

*191. Advertisement:

Coffee, Tea, or Vivarin?

Courtesy the J. B. Williams Company, Inc.

192. Advertisement:

Either you drive a Lynx, or you don't drive at all.

193. Advertisement:

Volvo: the car for people who think.

Courtesy Volvo, Inc.

194. Advertisement:

To lease or not lease. The answer is simple . . . to a professional.

195. Advertisement:

All my men wear English Leather or they wear nothing at all.

Courtesy Mem Company, Inc.

196. Sign on hardware store:

If we don't have it, you don't need it.

197. Advertisement:

If you're without it, you're not with it.

198. Overheard:

Everyone either goes to USC or wants to.

*199. Cliché:

Now or never

200. **Popular saying:**

There are only two kinds of people in the world: winners and losers.

*201. **George Orwell:**

Life is like that—take it or leave it.

*202. **Nietzsche:**

What does not destroy me makes me stronger.

*203. **Patrick Henry:**

Give me liberty or give me death.

204. **Old adage:**

All that is good in the world is either illegal, immoral, or fattening.

205. **Said of one's puppy:**

Then it's settled—we're going to stop spoiling her and start treating her like a dog.

206. **Senator Richard Russell:**

If we have to start over again with another Adam and Eve, I want them to be Americans and not Russians!

207. **To an interviewer:**

No one goes around asking questions without any motives, but if someone does ask questions without motives, that person must be insane, and why should anyone bother with someone insane?

208. **From the *New York Times* sports section, January 1, 1978:**

The wide world of football coaching is not round. It has two sides—win or lose, good or bad, sweet or sour, thumbs up or thumbs down.

209. **Editorial:**

Is the university becoming a place where researchers are teaching class rather than a place where teachers are doing research?

*210. **From a textbook:**

If a professor is competent, she will be successful as an educator even with dull students; and if she is incompetent, she will fail as an educator even with bright students. Hence it is irrelevant whether her students are dull or bright.

*211. **A variant of the above:**

Conscientious students will study even without the threat of impending examinations, and hopelessly lazy students won't study even with the threat of impending examinations. So examinations are useless as evidence for getting students to study.

212. **From Plato:**

Socrates: I know what you want to say, Meno. Do you realize what a debater's argument you are bringing up, that a man cannot search either for what he knows or for what he does not know? He cannot search for what he knows—since he knows it, there is no need to search—nor for what he does not know, for he does not know what to look for.

Meno: Does that argument not seem sound to you, Socrates?

Socrates: Not to me.

Meno: Can you tell me why?

Socrates: I can.

Plato's Meno (Stephanus 80d–81a), trans. G. M. A. Grube (Indianapolis, Ind.: Hackett, 1976), p. 13.

*213. **Letter to the editor:**

We've seen how communism leads to dictatorship. We have to avoid the Communist approach and stick to the free-market economy.

*214. **Editorial:**

The proposition before us is of utter and irreducible simplicity. Shall we preserve the ramparts of our freedom inviolate and pure or shall we succumb gradually but fully to the minions of the Communist enemy?

215. Newscaster:

It seems to me now more than ever before in our history, one is either for law enforcement or against it, either for mob rule or for the law, either loves a cop or hates him.

216. Letter to the editor:

If we don't kill people who will kill people, how many more people will they kill?

*217. Editorial:

We can become independent of Arab oil only by ruining our environment.

218. Letter to the editor:

We also must think of posterity and the environment. If we cannot enjoy the simple things in life like breathing fresh air and beholding nature's beauty, why live at all?

*219. Editorial:

The peoples of the world necessarily face widespread and horrible death within the next century. If enough food is produced to feed the expanding world population, then water and air will be polluted beyond tolerable limit by the amount of fertilizers and insecticides that are required. So people will die of water and air poisoning. If these fertilizers and insecticides are not used, not enough food will be able to be produced on the land available, and people will starve.

220. Letter to the editor:

There is not enough energy for both people and the war-making machine. A worldwide choice has to be made between the God of creation and the God of war and destruction.

221. Church circular:

Life is a battle between the flesh and spirit, between God and Mammon, between good and evil, between the forces of light and the forces of darkness. Christ says: "He who is not with me is against me."

*222. Letter to the editor:

God doesn't tolerate fence-riders in the cosmic sense. You must be either committed to Christ, or else fall in with the Devil. There are only two places to spend eternity: Heaven and Hell. You can't be somewhere in between.

*223. From J. B. Priestley, *Man and Time:*

Here I must be frank. There may exist a few superhumanly disinterested intellects, but I believe all the rest of us come down on one side of the fence or the other. In our secret depths, whenever we do our unspoken wishing, either we want life to be tidy, clear, fully understood, contained within definite limits, or we long for it to seem larger, wilder, stranger. Faced with some odd incident, either we wish to cut it down or build it up. On this level, below that of philosophies and rational opinions, either we reject or ignore the unknown, the apparently inexplicable, the marvelous and miraculous, or we welcome every sign of them. At one extreme is a narrow intolerant bigotry, snarling at anything outside the accepted world picture, and at the other is an idiotic credulity, the prey of any glib charlatan. At one end the world becomes a prison, at the other a madhouse....Yes, I would rather risk the madhouse than enter the prison.

©1964 Aldus Books, London, p. 194.

Begging the Question

Begging the question is the first of four fallacies arranged here under the subclass of *evading the facts*. The other three are *question-begging epithets, complex question,* and *special pleading.*

To beg the question is to assume (instead of prove) the point at issue. In its most elementary form it is simply a matter of repeating what was said (affirming something while implying, or believing, one had confirmed it). The following are some of the major forms it takes: (1) one argues A is so because it is (or A says so); (2) one argues A is so because of B (where B is the same as A); (3) one argues A is so because of B (where B is dependent on A); or (4) one argues A is so because of B (where B is even more suspect than A). In all these cases we are thrown back upon A—which we are asked to grant. Exercises 224 to 229 are examples of begging the question in form 1; exercises 230 to 250 are examples of begging the question in form 2; exercises 259 to 264 are examples of begging the question in form 3; and exercises 265 to 272 are examples of begging the question in form 4. There are, in addition, cases where one smuggles in an assumption that simply undercuts one's argument in even subtler ways. Exercises 273 to 276 are some examples of begging the question in this form. Fallacy, wit, and madness are closely related, and exercises 251 to 258 illustrate the presence and use of this device in the latter two cases.

224.

*225. From _Decision_:

How do we know that we have here in the Bible a right criterion of truth? We know because of the Bible's claims for itself. All through the Scripture are found frequent expressions such as "Thus says the Lord," "The Lord said," and "God spoke." Such statements occur no less than 1,904 times in the 39 books of the Old Testament.

Gilbert W. Kirby, "Is the Bible True?" _Decision,_ Vol. 15, No. 1, Jan. 1974, p. 4.
© 1974, The Billy Graham Association.

*226. A variant of the above:

We must believe what the prophet of the Mormons, Joseph Smith, says, for the Book of Mormon states that Smith is a true prophet; and whatever the Book of Mormon says comes directly from God, since Smith said that he received the book directly from an angel messenger from heaven.

*227. Overheard:

We can believe what it says in the college catalog because the catalog itself says it is the official publication of the college.

228. From a textbook:

Every statement in this book is true. And the authority for this is that the statement "Every statement in this book is true" is in fact a statement in this book.

229. From *National Lampoon:*

"Meat costs what it costs because that's what it costs. All those people do all those things. They all get paid and they all make a profit. If they didn't make a profit they wouldn't do what they do. And that would be bad."

May 1978, p. 88 (True section). From the U.S. Government pamphlet, *Mary Mutton and the Meat Group.*

*230. Remark:

Miracles are impossible, for they cannot happen.

*231. Overheard:

School isn't worthwhile because book learning doesn't pay off.

232. Overheard:

How do I know he is not guilty? Because he is innocent, that's why.

233. Overheard:

Poor people want to work because needy citizens desire to have employment.

*234. One student to another:

Of course hydrogen burns; it's combustible isn't it?

235. Remark:

It's obvious that we have free will since none of our acts are predetermined.

*236. Letter to the editor:

There's no question but that the deterioration in modern moral values is to be attributed to people's inability to distinguish what is right and what is wrong.

237. Advertisement on a can:

TOM'S PEAS AND CARROTS

IN

BUTTER SAUCE

containing

butter sauce, peas and carrots

238. Editorial:

This measure is designed to reduce the staggering debt of the nation, for it proposes to pay off some of that debt.

*239 Advertisement:

Our low prices are the direct result of our lowered price policy.

240. Francis Bacon:

He that hath wife and children hath given hostages to fortune; for they are impediments to great enterprise, either of virtue or mischief.

241. Questioned by his friend Horatio what the ghost said, Hamlet replies (act 1, scene 5):

There's ne'er a villain dwelling in all Denmark. But he's an arrant knave.

*242. Editorial:

All people should be free, for liberty is the universal right of humanity.

*243. Letter to the editor:

Death for traitors is properly justified because it is right to put to death those who betray our country.

244. **Editorial:**

The bill before the House is well calculated to elevate the character of education in the country, for the general standard of our schools will be raised by it.

245. **Popular maxim:**

Everyone has a price, for money is a standard of value all recognize.

*246. **Letter:**

Indiscriminate charity is reprehensible because everyone would agree it is wrong not to distinguish between deserving and undeserving cases.

247. **From a sermon:**

The human soul is immortal, for not a shadow of a doubt exists that the noble spirit that animates the human clay will survive the catastrophe of death and continue to exist for all eternity.

248. **Letter:**

Popular government results in corruption, inefficiency, and chaos, for popular government is run by all the people and when you get all the people into government, you have one that rests only upon popular consent. You always find that when you have a government of that kind it is never honest, efficient, or orderly.

249. **From *Time* magazine:**

How to account for all this added heft? Laments Statistician Sidney Abraham, the main author of the center's report: "Adults are obviously not getting appreciably taller, and they usually do not get more muscular. All we can say is that the weight increase we found is due to fat."

"Land of the Fat: America's Weight Woes Grow," January 2, 1978, p. 53. Reprinted by permission from *Time, The Weekly Newsmagazine:* Copyright Time Inc. 1978.

*250. Example from a classic logic text:

To allow every man an unbounded freedom of speech must always be, on the whole, advantageous to the State; for it is highly conducive to the interests of the community, that each individual should enjoy a liberty perfectly unlimited, of expressing his sentiments.

Richard Whately, *Elements of Logic* (London: John W. Parker, 1850), p. 134.

*251. Famous remark by Joe E. Lewis:

I don't want to be a millionaire; I just want to live like one!

252. Famous interchange:

F. Scott Fitzgerald: The very rich are different from you and me.
Ernest Hemingway: Yes, they have more money.

See Hemingway's oblique reference to this in *The Snows of Kilimanjaro* (New York: Modern Library, 1938), p. 170.

*253. Overheard:

Question: How can you tell the age of a snake?
Answer: It is extremely difficult to tell the age of a snake unless you know exactly when it was born.

*254. From Harris:

Husband: Dear, where are my cuff links?
Wife: Where you left them!

Thomas A. Harris, *I'm OK—You're OK* (New York: Avon Books, 1973), pp. 105–106.

255. Overheard:

Doctor: You know none of us is getting younger.
Patient: I don't care about not getting younger; I just don't want to get older!

256. In Molière's play *Le Medicin Volant* (1660), scene 3, a character who is impersonating a doctor tells a gullible client:

Hippocrates has said—and Galen confirms it with many persuasive arguments—that when a girl is not in good health she is sick.

One Act Comedies of Molière, trans. Albert Bermel (New York: World Publishing, 1964), p. 33.

*257. Overheard:

Husband: I'm not jealous; I just don't like her speaking to everyone!

258. The clown in *Hamlet* (act 5, scene 1) upon being asked how the prince had gone mad, replies:

By losing his wits.

*259. Overheard:

Dad: The liberals don't like nuclear plants.
Son: Why is that?
Dad: Because it is their philosophy.
Son: What is their philosophy?
Dad: Not to like nuclear plants!

*260. Overheard:

Fan: The Broncos are a better team than the Cowboys.
Friend: Why?
Fan: Because they've won more games.
Friend: And why have they won more games?
Fan: Because they're a better team!

261. Overheard:

Bill: I enjoy reading only good books.
Tom: How do you know when they're good?
Bill: If they're not good, I don't enjoy them.

262. From a logic text:

What makes you think Lefty is a crook?
Well, look at the crooks he associates with!
How do you know they are crooks?
Well, anyone who'd associate with a crook like Lefty....

*263. Variant of the above:

> *He talks with angels.*
> *How do you know?*
> *He said he did.*
> *But suppose that he lied!*
> *O, perish the thought! How could any man who is*
> *capable of talking with angels lie!*

264.   From *The Little Prince* by Antoine de Saint Exupery:

The next planet was inhabited by a tippler. This was a very short visit, but it plunged the little prince into deep dejection.

"What are you doing there?" he said to the tippler, whom he found settled down in silence before a collection of empty bottles and also a collection of full bottles.

"I am drinking," replied the tippler, with a lugubrious air.

"Why are you drinking?" demanded the little prince.

"So that I may forget," replied the tippler.

"Forget what?" inquired the little prince, who already was sorry for him.

"Forget that I am ashamed," the tippler confessed, hanging his head.

"Ashamed of what?" insisted the little prince, who wanted to help him.

"Ashamed of drinking!" The tippler brought his speech to an end, and shut himself up in an impregnable silence.

And the little prince went away, puzzled.

"The grown-ups are certainly very, very odd," he said to himself, as he continued on his journey.

Many editions. The above taken from the edition translated from the French by Katherine Woods. New York: Harcourt, Brace & World, Inc., 1971, (Chapter 12) and William Heinemann, Ltd., pp. 50–52.

265.   Letter to the editor:

These men are traitors because they are opposed to the war, and opposition to the war is an opposition to the government, and opposition to the government is traitorous.

*266. Philosophy essay:

Everything that happens has a cause; for if something should occur without a cause, it would have been caused by itself. But this is impossible.

267. **Overheard:**

I don't care why Mary stayed over at Mark's. I don't expect my daughter, and I think a lady at that, to ever spend the night at a boy's house!

268. **From a pamphlet:**

Communism is the best form of government because it alone takes care of the interests of the common people.

269. **Overheard:**

The crime this man committed is the result of his childhood environment; for all such crimes are rooted in childhood environment, as this man's case proves.

*270. **From a philosophy essay:**

Reality in itself must be as it appears to the five senses; for if it were not, then there would be no other way that we could know it.

271. **Summary of a contemporary philosophical position:**

Moral beliefs are unjustifiable because they are not verifiable in sense experience.

*272. **From Rabi, 1944 Nobel laureate in physics:**

What is time? This question has intrigued mystics and philosophers for centuries. The real answer was given only in this century by Einstein, who said, in effect, that time is simply what a clock reads.

Isidor Isaac Rabi, "Introduction," *Time Life Science Library* (New York: Time Inc., 1966), p. 7.

273. **Overheard:**

Smith cannot have told you a lie when she said she was my cousin, for no cousin of mine would ever tell a lie.

274. **To a clerk at a motel:**

Certainly she's my wife. I'm her husband.

*275. Student essay:

Everything in the world has a cause: since if it did not we should have effects without causes.

*276. From Howard Pospesel:

"In the early fifties a prospector named James Kidd vanished in Arizona, leaving a handwritten will directing that his fortune ($230,000) go into a research or some scientific proof of a soul of the human body which leaves at death. One hundred and thirty individuals and institutions filed for Kidd's estate, including Dr. Richard Spurney, a junior college teacher. Spurney submitted to the court a foot-thick pile of 'evidence' which included three unpublished books. He had fifty 'proofs' of the existence of the soul. Spurney summarized his main proof as follows:

> Death is decomposition. Hence, what cannot decompose cannot die. But decomposition requires divisibility into parts. Thus what is not divisible into parts cannot die. But divisibility into parts requires matter. Hence what has no matter in it is not divisible into parts and so cannot decompose, and so is necessarily immortal.

"...After hearing testimony for thirteen weeks, the court awarded the estate to the Barrow Neurological Institute of Phoenix, Arizona."

Howard Pospesel, *Arguments: Deductive Logic Exercises* (Englewood Cliffs, N.J.: Prentice-Hall, 1971), p. 19.

Question-Begging Epithets

We can beg the question at issue with a single word—an *epithet,* as some writers call it. This is a descriptive label or tag which we attach to a person, a thing, or an idea in an effort to condemn it out-of-hand. Marshall McLuhan once referred to the tactic as "libel by label." In the context of political debates it is a matter of "mudslinging." (It was in this connection that Adlai Stevenson once remarked that "he who slings mud loses ground.") Some other names under which it is sometimes described are *name calling, loaded words, controversial phrases, verbal suggestion,* and *emotive language.*

Since we can beg the question not only by using uncomplimentary (*dyslogistic*) terms but also by using complimentary (*eulogistic*) terms, several examples of the latter type have also been included.

Such epithets are objectionable because they assume attitudes of approval or disapproval without providing evidence that such attitudes are justified. Rather than simply providing us with the relevant facts and allowing us to judge for ourselves what our attitude to them should be, they try to do this for us.

Exercises 277 to 289 are examples of question-begging epithets.

*277. Speaker:

This is a deliberate plot against the American people.

278. Witness:

The scoundrel hounded his wife to the grave.

*279. Editorial:

This measure is calculated to subvert the just aspirations of our people.

*280. Editorial:

This is a proposal that is offensive to every enlightened conscience.

281. Bumper sticker:

LIBERTARIAN PARTY:
Don't rob.
It's a crime to compete with Government.

*282. From Gross:

James Simon Kunen, a young heavily haired dissident, has become a wealthy post adolescent as author of the best selling *The Strawberry Statement*.

Milton Gross, *Chicago Sun Times*, August 2, 1970. Reprinted with permission from *The Chicago Sun-Times*.

ᵗ283. Letter to the editor:

A man should find it degrading to live on a dole or any payments made to him without his being required to render some service in return. But how many of them do feel degraded by it. From an economic standpoint such loafers are simply parasites and should be dealt with accordingly.

*284.   Letter:

May I have 3 minutes of your time?

Because you are one of the National Right to Work Foundation's best friends, I wanted you to have a look at the enclosed letter. I knew you would be interested in the outrageous injustice described here.

It's a letter from The Throneberry Family...hard-working, honest people threatened by Big Union Goons.

The National Right to Work Foundation is trying to help the Throneberry family. Unless we stop the union goons now, who knows who the next victim will be?

When the union went on strike against Stanley Tool Company in Shelbyville, Tennessee, a nightmare began for the Throneberry family.

Unfortunately, that nightmare...the union violence aimed at the Throneberrys...still continues. But why are the union thugs attacking this family?

Because Mrs. Throneberry had the courage to apply for a job at Stanley Tool during the strike.

During the 34 weeks of the strike, physical violence and destruction of property became the rule in Shelbyville, not the exception.

The strike divided the small Tennessee town between militant unionists and law-abiding citizens who wanted to work, like Mrs. Throneberry.

Signed Thomas J. Harris, Chairman

Courtesy National Right to Work Legal Defense Foundation Inc.

*285.   Editorial:

Sen. Charles H. Percy (R.-Ill.) should be lauded for submitting legislation to repeal the no-knock statute of the Drug Abuse Prevention and Control Act of 1970. Unannounced destructive raids perpetrated by government agents rabid for the "bust" are at best the antithesis of law, order, and democracy. Drug trafficking has been and continues to be a menace to the welfare of America's youth; yet Gestapo tactics that encroach upon the innocent cannot be condoned.

286. "America's War on Drugs: the Moral Equivalent of the Inquisition" by Thomas Szasz:

What is behind this fateful moral and political transformation, which has resulted in the rejection by the overwhelming majority of Americans of their right to self-control over their diets and drugs in favor of the alleged protection of their health from their own actions by a medically corrupt and corrupted state? How could it have come about in view of the obvious parallels between the freedom to put things into one's mind and its restriction by the state by means of censorship of the press, and the freedom to put things into one's body and its restriction by the state by means of drug controls?

Los Angeles Times, January 8, 1978.

*287. Letter to the editor:

This former Marine Corps master sergeant, with voluntary service in two wars, felt nothing for the financial plight of our little vacationing soldier boys in Germany. The U.S. military is totally overpaid and pampered. I've had occasions to observe our soldiers' behavior in Germany, both married and unmarried: and in most cases, if not plain ignorant, it's shabby. Many Germans are totally fed up with the ill-disciplined, prideless people we call the military.

Time Magazine, January 16, 1978, p. 7. Reprinted by permission from *Time, The Weekly Newsmagazine;* Copyright Time Inc. 1978.

288. When Carl Sandburg's poem "Chicago" first appeared in the March 1914 issue of *Poetry Magazine,* the editor of *Dial* wrote:

The typographical arrangement for this jargon creates a suspicion that it is intended to be taken as some form of poetry, and the suspicion is confirmed by the fact that it stands in the forefront of the latest issue of a futile little periodical described as 'a magazine of verse.' ... We think that such an effusion as the one now under consideration is nothing less than an impudent affront to the poetry-loving public.

"New Lamps for Old." *The Dial,* Vol. 56, No. 666, March 16, 1914, pp. 231–32.

*289. Senator Robert C. Byrd (D.-W. Vir.) in a statement to the President's Commission on Campus Unrest, September 9, 1970:

It is high time that professors who distort the perspective of young minds, who advocate the overthrow of our system of government, who corrupt and pervert the educational process, be purged from our educational institutions.

The social studies in particular are dominated by the high priests of radicalism, and it is little wonder that many American young people get a badly distorted picture of their country, its present and its past.

There is something dreadfully wrong with college governing boards and administrations which allow faculties to become overloaded with fuzzy-minded, phony liberals whose heroes are Che Guevara, Fidel Castro, Ho Chi Minh, and Mao Tse-Tung.

Complex Question

The fallacy of the *complex question* arises when a question is phrased so that it cannot be answered without granting a particular answer to some question at issue. The old saw "have you stopped beating your wife yet?" is still probably the clearest example of this fallacy. Like the fallacy of question-begging epithets, it goes under a variety of names, such as *loaded question, trick question, leading question, false question,* and the *fallacy of many questions.* Exercises 290 to 330 are examples of this fallacy.

Of these, exposing the hidden assumption is all that is required in exercises 290 to 294; exercises 295 to 298 can best be handled by dividing the question; exercises 299 to 310 combine complex question with question-begging epithets; exercises 311 to 314 ask for explanations for "facts" that are either untrue or not yet established; exercises 315 to 329 are examples of the exploitation of this device in advertising and sales; and exercise 330 is an example of the device in interpersonal relations.

Regarding the use of this device in advertising and sales, it might be interesting to observe that in exercises 314 to 320 the device used is to assume on our behalf the positive merits of the product promoted; in exercises 321 to 323 an attempt is made to help us make up our minds to buy the product (long before we have decided we even want it); in exercises 324 to 329 an attempt is made to get us to buy the product in question by playing on our fears.

290. **From the nineteenth-century German philosopher Nietzsche:**

Why do I know more than other people? Why, in general, am I so clever?

Friedrich Nietzsche, "Ecce Homo," in *The Philosophy of Nietzsche,* trans. Clifton Fadiman (New York: Random House, 1954), p. 833.

291. **Editorial:**

Could the reason for the unexcelled superiority of American optical goods be this: they have been made under a competitive system?

292. **Editorial:**

Why are we not urgently exploring alternate energy sources or mandating that new cars cannot be built that do not get significantly better mileage?

*293. **Overheard:**

Why was not the new building given to the school?

294. **Overheard:**

Why wasn't I told?

295. **Overheard:**

Are you for the Republicans and prosperity?

296. **Couple:**

You say we ought to discuss whether or not to buy a new car now. All right, I agree. Let's discuss the matter. Which should we get, a Ford or a Chevy?

*297. **Editorial:**

Is this policy going to lead to ruinous deflation?

298. **Speaker:**

Shall we favor extending federal aid to parochial schools and thereby violate the First Amendment of the Constitution regarding the separation of church and state?

299. Flyer:

Should government rob its citizens?

300. Editorial:

Why do conservatives give all-out and unthinking support to a shaky technology propped up by government subsidies?

The Spotlight. Published by Liberty Lobby. July 16, 1979, p. 27.

*301. Parent:

How much longer are you going to waste your time in school when you might be doing a man's work in the world and contributing to society? If you had any sense of social responsibility, you would leave immediately.

302. Jimmy Carter's 1976 presidential campaign slogan:

For America's third century, why not the best?

303. Letter to the editor:

Don't the President and his advisers see that the Turks do not want American arms for their defense against Soviet Russia but for aggression against Greece?

Los Angeles Times, May 23, 1978, Part II.

*304. Overheard:

Why is it now that every high school has its full-to-the-gills student parking lot? Why is it that every teenager regards his or her own car as a birthright?

305. Letter:

Are these recalcitrant school failures to continue to be allowed to endanger, disrupt, and destroy the academic opportunities of those serious students who desire learning the requirements of our complicated technological and multifaceted culture?

*306. Interviewer:

Senator, why is it so hard for you to come to a conclusion?

307. Religious pamphlet:

Is the School House the Proper Place to Teach Raw Sex?

308. Overheard:

Do the people in your country still cheat visitors?

309. Overheard:

How much longer is Pat going to pretend to know everything?

310. From *Rolling Stone Magazine,* December 1977:

Has Rod Stewart finally made an album worth listening to?

Rolling Stone by Straight Arrow Publishers, Inc. © 1977. All Rights Reserved.
Reprinted by permission.

311. Old riddle:

What is the biggest number?

*312. Old riddle:

What time was it before time began?

313. Advertisement:

Can we know our past lives?

314. Advertisement for computerized horoscope:

Can the stars improve your love life?

*315. Sugar advertisement:

If sugar is so fattening, how come so many kids are thin?

*316. Cosmetic advertisement:

Are your lashes as beautiful as they used to be?

317. Soap advertisement:

Aren't you glad you use Dial?

Courtesy Armour Dial, Inc.

318. Commercial:

Why pay more when you can shop at Vons?

Courtesy Vons.

319. Advertisement for Accutron Watch by Bulova:

Do you have an unfaithful watch?

Courtesy Bulova Watch Co. (from a 1972 ad campaign).

320. Another advertisement for the same watch:

Is the old ticker running out on you?

Courtesy Bulova Watch Co. (from a 1972 ad campaign).

*321. Sales representative:

Would you like me to make this contract out for twelve or eighteen months?

322. Another sales representative:

We make our own deliveries. What days would be best to deliver this item to you?

323. Door-to-door salesman:

Good morning, madam. Thank you for answering my knock on your door. Allow me to introduce myself: I'm Al Cooper; I'm selling vacuum cleaners. What kind of accessories do you want with the one you're going to buy?

324. Advertisement:

Haven't you promised yourself a Cadillac long enough?

Courtesy of Cadillac Motor Car Division, General Motors Corp.

*325. Fountain pen advertisement:

How many other lifetime investments cost as little as $13.50?

326. Advertisement for English muffins:

Doesn't your family deserve the best?

*327. Airline advertisement:

How to get to Tokyo without being worn out for once?

328. Commercial:

Is your car entirely safe? Install Sure-Grip Tires today!

*329. Commercial:

Would your wife and children have security if you died today? If not shouldn't you buy more insurance?

*330. From Thomas Harris:

Husband: Where did you hide the can opener?
Wife: I hid it next to the tablespoons, darling.

Thomas A. Harris, *I'm OK–You're OK* (New York: Avon Books, 1973), pp. 114–16.

Special Pleading

When we begin to think of ourselves as being special and therefore subject to a different standard than one we are inclined to apply to others, we fall victim to the fallacy of *special pleading*. Bertrand Russell once described this characteristic as a matter of the way some people "conjugate" certain words—"firmness" for example: "I am firm; you are stubborn; he is pig-headed." Without using this particular label, Edwin Newman, in his book *Strictly Speaking*, provides a striking and illuminating example of this tactic:

> The war in Indochina produced a host of terms that media folks accepted at their peril: protective reaction strike, surgical bombing, free-fire zone, interdiction, contingency capability, New Life Hamlet—which in sterner days was a refugee camp—and many more. Money paid to the family of a South Vietnamese civilian killed by mistake was a condolence award.
>
> In February, 1971, South Vietnamese forces, with Ameri-

can air support, moved into Laos. Rarely had the importance the government attached to language been made so clear. An incursion, Washington called it, and there were official objections to our calling it an invasion, evidently in the belief that incursion implied something softer than invasion did, and that an incursion was permissible where perhaps an invasion was not.

At this point, media interest led to the dictionary, where an incursionist was defined as an invader, incursionary as invading, and incursion as entering into territory with hostile intent; a sudden invasion. . . .

This being the case, why not say incursion and give the government its heart's desire? Because calling it an incursion was a public relations exercise, an attempt to make it appear less grim than it was and acceptable to the American people.

The distinction between incursion and invasion was a distinction without a difference, in grammar and in fact. The incursion into, or invasion of, Cambodia in 1970 enormously increased death and destruction there, and the incursion into, or invasion of, Laos increased death and destruction there. It is not the business of news people to exaggerate any of this, but *it* is not their business to water it down either.

*331. From Thomas Szasz:

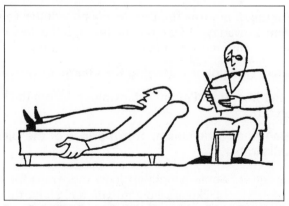

". . . the patient is narcissistic, the analyst has self-esteem; the patient is inhibited, the analyst has self-control; the patient is promiscuous, the analyst is liberated. . . ."

332. Overheard:

I know how to be thrifty, but Larry is a simple tightwad.

333. Overheard:

Did you see the way the Bruins savagely attacked our players? Our team, on the other hand, played with great spirit and enthusiasm.

334.

"The Wizard of Id" by permission of Johnny Hart and Field Enterprises, Inc.

335. Member of a sorority:

I might be a little overweight, but the rest of the girls in my house are a lot fatter than me.

*336. Overheard:

Speaking of not trusting people, it's no wonder you can't trust anyone nowadays. I was looking through the desk of one of my roomers, and you won't believe what I found.

337. Woman (indignantly denying the charge of hoarding):

I'm not hoarding. I'm only stocking up before the hoarders get everything.

*338. Sultan Khaled Hethelem, crown prince of the Ujman tribe of Saudi Arabia:

Interviewer: Some people criticize the Arabs for using oil to blackmail; how do you feel about this?

Sultan: We are not trying to blackmail Western Europe or Japan; we are just trying to convince them to help us.

Rufii, International Relations Newsletter, University of Southern California, Vol. 11, No. 2.

339. **From *The Invisible Scar* by Caroline Bird:**

Many Americans with money took fright. They turned in their dollars for gold and shipped it out of the country without the slightest feeling of disloyalty, at the same time denouncing ordinary depositors, who questioned the soundness of their banks, as "unpatriotic."

(New York: David McKay, Co., 1966). Copyright © 1966 by Caroline Bird. Reprinted with permission of Longman Inc. and author.

340. **Commercial:**

Phillips 66. The Performance Company.

Courtesy Phillips Petroleum Company.

341. **News report:**

At the center for the Study of American Business, which he heads at Washington University in St. Louis, Economist Weidenbaum, 51, is examining how the policies and regulations of big, intrusive Government are feeding inflation, impeding efficiency and otherwise rubbing up against private citizens. Given the bullish bulge of bureaucratic power, his institute is quite a growth enterprise.

"Battling the B.I.G. Bulge," *Time,* May 28, 1978, p. 76. Reprinted by permission from *Time, The Weekly Newsmagazine;* Copyright Time Inc. 1978.

342. **From a speech delivered by Benito Mussolini in Rome, June 10, 1940:**

This gigantic conflict is only a phase of the logical development of our revolution. It is the conflict of poor, numerous peoples who labor against starvers who ferociously cling to a monopoly of all riches and all gold on earth. It is a conflict of fruitful, useful people against peoples who are in decline.

343. **From a speech by Andrei Y. Vishinsky delivered in the U.N. General Assembly September 20, 1950:**

While in those countries which have entered the North Atlantic alliance a mad armament race is taking place and an unbridled war propaganda is being broadcast, and while the war psychosis is being incited more and more, the Soviet Union is the scene of peaceful, creative work. All the forces of our country are directed to the fostering of our national economy and to improving the standard of living and the welfare of the Soviet people.

344.   From Mao Tse-tung:

Riding roughshod everywhere, U.S. imperialism has made itself the enemy of the people of the world and has increasingly isolated itself. Those who refuse to be enslaved will never be cowed by the atom bombs and hydrogen bombs in the hands of the U.S. imperialists. The raging tide of the people of the world against the U.S. aggressors is irresistible. Their struggle against U.S. imperialism and its lackeys will assuredly win still greater victories.

Mao Tse-tung, "Statement Expressing the Chinese People's Firm Support for the Panamanian People's Just Patriotic Struggle," in *Statements by Mao-Tse tung* (Peking: Foreign Languages Press, 1964), pp. 11–13.

*345.   News report:

It was exactly ten minutes after seven on Thursday evening, June 12, that Mrs. Lela Bass, 73, stood combing her long gray hair in the backyard of her white frame house in Port Neches, Texas. Casually she turned and saw for the first time an eerie outline etched in the plastic of her backdoor screen: a bearded, long-haired man with a halo, looking east toward a fig tree in the yard. It was, she was certain, Jesus Christ. Neighbors spread the word, and since then, more than 50,000 curious visitors have descended on the Bass home to share her vision.

Port Neches is a bleak Gulf Coast industrial town that is also intensely religious. On Sundays, most of its 10,000 inhabitants troop loyally to one or another of the town's 35 churches; some have so much fundamentalist fear of the Lord that they respectfully refer to Jesus as "Mr. Christ." The shared excitement over the phenomenon has brought blacks and whites together in a proximity unusual for Port Neches; but the two races sometimes differ on what they see. One white farmer, who claims that he has taken some 25 photographs showing images of "the Christ Child, the Virgin Mary, the Three Wise Men, and angels," scoffed at Negro viewers. "These blacks come away saying they've seen Martin Luther King, Bobby Kennedy and J.F.K. Boy, those people sure have an imagination."

"Visions: The Image of Mr. Christ." *Time,* July 11, 1969, p. 48. Reprinted by permission from *Time, The Weekly Newsmagazine;* Copyright Time Inc. 1969.

346. News report:

Settlers in Santa Fe were in a festive mood when they chose the city plaza as the site of a 33-ft. obelisk dedicated to "the heroes who have fallen in the various battles with savage Indians in the Territory of New Mexico." A plaque bearing that inscription went onto the monument's cornerstone in 1868, and there is no record that anyone found it objectionable. Times change, however, and leaders of the American Indian Movement now condemn use of the word savage as "racism and prejudice." They want the obelisk destroyed. Governor Bruce King agreed with their sentiments but not with their draconian solution, so he suggested substituting *fierce* for *savage*. "That's playing with history," objected State Cultural Properties Review Committee Chairman Albert Schroeder. "It's the most ridiculous thing I've ever heard," said State Historian Myra Ellen Jenkins. As the controversy wore on, one Santa Fe wag suggested placing a fig leaf over the word savage. Someone else proposed a second obelisk dedicated to "the gallant Indians who died fighting for their homeland." The Santa Fe city council turned the problem over to the state. Its solution: an explanatory plaque, to be placed alongside the offending monument, that will read in part: "Monument texts are wont to reflect the character of the time in which they are written and the temper of those who wrote them. Hopefully, attitudes change and prejudices dissolve."

"Revisionist History," *Time*, September 17, 1973, p. 12. Reprinted by permission from *Time, The Weekly Newsmagazine;* Copyright Time Inc. 1973.

*347. From Yasir Arafat, PLO leader:

The United Nations charter gives all peoples the right to self-determination. Not only are we deprived of this right, but of our country itself. So every Palestinian, wherever he is, is a victim of the same injustice practiced against him. This feeling was the common motive for all Palestinians to regain their identity, their nationality, and their homeland through armed struggle. Since we are facing an enemy supported by imperialist funds and technical achievements, we have chosen long-term war as the strategy for our struggle.

Yasir Arafat, in *Terror and Urban Guerrillas: A Study of Tactics and Documents*, ed. Jay Mallin (Coral Gables, Fla.: University of Miami Press, 1971), p. 47.

*348. Some articles of the Palestinian National Covenant, as adopted by the Palestine National Council in Cairo in 1968:

Article 5) The Palestinians are the Arab citizens who were living permanently in Palestine until 1947, whether they were expelled from there or remained. Whoever is born to a Palestinian Arab father after this date, within Palestine or outside it, is a Palestinian.

Article 6) Jews who were living permanently in Palestine until the beginning of the Zionist invasion [defined in the resolution of the Congress as having begun in 1917] will be considered Palestinians.

Article 15) The liberation of Palestine, from an Arab viewpoint, is a national duty to repulse the Zionist, Imperialist invasion from the great Arab homeland and to purge the Zionist presence from Palestine....

From Lester A. Sobel, editor, *Palestinian Impassee: Arab Guerrillas and International Terror.* New York, N.Y.: Facts on File, Inc., 1977, pp. 23–24.

False Analogy

False analogy is the first of three fallacies arranged here under the subclass *distorting the facts,* the remaining two being *false cause* and *irrelevant thesis.*

An analogy is a likeness between one thing and another (since A is like B in certain ways, then what is true of A must be true of B). Although analogies never prove anything, they can be extremely useful in making difficult ideas clear. To be useful, however, we have to make sure whenever we use an analogy that the things being compared resemble each other in significant ways and differ from each other in insignificant ways. If it is the other way around (the two things being compared resemble each other only in trifling ways and differ in important ones), we are likely to be led astray by the analogy used. Exercises 349 to 373 are examples of false analogy.

Of these exercises, 350 to 358 are short and classic examples of the fallacy, some designed to show how an intriguing metaphor may be behind the deception; exercises 359 to 361 are a few absurd examples of the fallacy; and the remaining exercises (from 362 to 373) are, again, classic examples of the fallacy, arranged in order of growing difficulty.

349. **Early in 1978 Wendy Joy Helander filed a $9 million civil suit in New York City, alleging forcible abduction and conspiracy. The abduction was part of an eighty-six-day "deprogram-**

ming" effort to get her away from the Rev. Sun Myung Moon's Unification Church. Ms. Helander's suit was based on her second alleged "kidnapping" by her parents and Mr. Ted Patrick. Following the first, which occurred in 1975, she obtained a $5,000 judgment against Mr. Ted Patrick. Her parents' countersuit alleged that their "rescue attempts" together with court costs now exceeded $40,000. The cartoon below is a commentary on this and other cases like it.

*350. Politician:

A nuclear power plant is infinitely safer than eating, because three hundred people choke to death on food every year.

351. Letter to the editor:

Stupid! Stupid! Stupid! Refusing to let Edison build a safe, clean plant at this time in history—because "it will destroy scenic beauty" and "might hurt the fish"—is like refusing to get an iron lung to keep you alive because it will destroy the decor of your bedroom and might fall on the kids! The Coastline Commission must have been suffering a power shortage between their ears when they made this decision. To sum up: Dimwits create blackouts. Pray for an appeal.

*352. Popular saying:

The razor edge of his intellect will be blunted by constant use.

353. King James I of England:

If you cut off the head of a body, the other organs cannot function and the body dies. Similarly, if you cut off the head of the state, the state may flop around awhile, but it is due to perish in time or become easy prey to its neighbors.

*354. From a review:

From reading Thomas Thompson's vivid and terrifying book, *Hearts,* I knew that for some years two world-famous Houston surgeons, Doctors Cooley and Michael DeBakey, had been engaged like rival Aztec kings in a struggle to see who could tear open the most hearts.

355. From the nineteenth-century German philosopher Hegel:

War has the deep meaning that by it the ethical health of the nation is preserved and their finite aims uprooted. And as the winds which sweep over the ocean prevent the decay that would result from its perpetual calm, so war protects the people from the corruption which an everlasting peace would bring upon it.

G. F. W. Hegel, *The Philosophy of Right* (1821), part 3, section 3.

356. From an astrology circular:

People who dislike astrology will sometimes say, out of malice, that the stars' distances are sufficient to make the effects of their vibrations negligible. This is easily refuted. First, the space between earth and the stars is nearly a vacuum, empty enough to let vibrations pass almost undiminished in vigor. Second, the Chicago exposition in 1933 was opened by a ray of light from the star Arcturus hitting a photoelectric cell. If a star as remote as Arcturus can open a fair in Illinois, it is very careless to argue that the stars are too far away to affect our lives.

*357. From a logic textbook:

Why should we criticize and punish human beings for their

actions? Whatever they do is an expression of their nature, and they cannot help it. Are we angry with the stone for falling, and the flame for rising?

*358. From *Alive:*

"You cannot condemn what they did," said Monsignor Andrés Rubio, Auxiliary Bishop of Montevideo, "when it was the only possibility of survival.... Eating someone who has died in order to survive is incorporating their substance, and it is quite possible to compare this with a graft. Flesh survives when assimilated by someone in extreme need, just as it does when an eye or heart of a dead man is grafted onto a living man.... What would we have done in a similar situation?"

Piers Paul Read, *Alive: The Story of the Andes Survivors* (Philadelphia: Lippincott, 1974), p. 340. Copyright © 1974 by Piers Paul Read.

359. Sitcom dialogue:

Jeff: That stuff's not my regular beer!
Sue: That beer was on sale, so I bought it!
Jeff: If dirty socks were on sale, would you buy them!

*360. News report:

CONCORD, N.C.—A married woman in her late twenties says she has been going topless for the last four years when the weather is pleasant, when she is working in her yard, driving her car, or riding a motorcycle with her husband.... A state highway patrolman, T. L. Hooks, stopped her Sunday while she was riding topless on a motorcycle with her husband.... Hooks said he later let her go because there is no law prohibiting her from being topless in public. "I guess it's not legally indecent to do that," he said, "but I still believe it's improper. It could cause accidents." The woman's husband supports her action. "You can't have two sets of moral values, one for men and the other for women," he said. And she says: "If a man can go without a shirt, then so can I. There's not much difference between the chest of a man and the chest of a woman, A little more fat on the woman, a little more hair on the man...."

" 'If a Man Can Go Without a Shirt, then So Can I' " (Associated Press), *Miami News,* July 25, 1974. (Taken from Howard Pospesel and David Marans, *Arguments: Deductive Logic Exercises*, 2nd ed. (Englewood Cliffs, N.J.: Prentice-Hall, 1978), p. 48.

361. **Huck Finn is trying to explain to Jim that there is no absolutely correct way of speaking. The Frenchman's "Polly-voo-franzy?" means the same thing as our "Do you speak French?" To this Jim replies:**

"Well, den, why couldn't he say it?"

"Why, he *is* a-saying it. That's a Frenchman's *way* of saying it."

"Well, it's a blame ridicklous way en I doan' want to hear no mo' 'bout it. Dey ain' no sense in it."

"Looky here, Jim; does a cat talk like we do?"

"No, a cat don't."

"Well, does a cow?"

"No, a cow don't, nuther."

"Does a cat talk like a cow, or a cow talk like a cat?"

"No, dey don't."

"It's natural and right for 'em to talk different from each other, ain't it?"

"Course."

"And ain't it natural and right for a cat and a cow to talk different from *us*?"

"Why, mos' sholy it is."

"Well, then, why ain't it natural and right for a *Frenchman* to talk different from us? You answer me that."

"Is a cat a man, Huck?"

"No."

"Well, den, dey ain't no sense in a cat talkin' like a man. Is a cow a man?—or is a cow a cat?"

"No, she ain't either of them."

"Well, den, she ain't got no business to talk like either one er the yuther of 'em. Is a Frenchman a man?"

"Yes."

"Well, den! Dad blame it, why doan he *talk* like a man? You answer me dat!"

Samuel Langhorne Clemens, *The Adventures of Huckleberry Finn* (New York: Harper & Row, Pub., 1951). pp. 86–87.

362. **Overheard:**

The highway patrol should set up roadblocks all over the city on the weekends and stop all cars to check for drunk drivers. After all, the highway patrol does precisely this during the Christmas and New Year holidays, and many dangerous drivers are thereby apprehended.

363. From a logic textbook:

Absolute music is perhaps the most eloquent and moving form of art, although it tells no story. Abstract painting and sculpture are among the most magnificent products of human creativity, although neither of them has any story to tell. Therefore the story it contains contributes nothing to the excellence of a novel or a drama as a work of art.

364. Letter to the editor:

It is unjust and unfair to deny to a man the right to hold political office on account of his racial origin; it is unjust and unfair to deny to a woman the right to higher education on account of her sex; it is unjust and unfair to tax railroad trains and let trucks and buses travel free over public roads; it is unjust and unfair for politicians to cater to rich districts and to ignore poor ones; all these cases show that no injustice or unfairness is to be permitted on account of accidents of circumstance. Hence, certainly, it is unjust and unfair to keep my boy Bobby out of the Westwood troop of the Cub Scouts just because he lives in North Hollywood.

365. From a dialogue in We:

"It is absurd because there can be no revolution. Because our revolution was the final one. And there can be no others. Everyone knows this...."

"My dear—you are a mathematician. More—you are a philosopher, a mathematical philosopher. Well, then: name me the final number."

"What do you mean? I...I don't understand: what final number?"

"Well, the final, the ultimate, the largest."

"But that's preposterous! If the number of numbers is infinite, how can there be a final number?"

"Then how can there be a final revolution?"

Yevgeny Zamyatin, We (New York: Viking, 1972), p. 152.

*366. A reporter who pressed Robert Welch, leader of the John Birch Society, to name the Communist conspirators in Washington, reported the following answer:

Suppose I'm walking on a golf course with the president of the club and we see divots all along the way. I say to him "You've got some members who don't care about the course." He asks me to name them. I tell him, "I can't name them, but they must be here."

James Wechsler, New York Post, October 11, 1965. Reprinted by permission of the New York Post. © 1965, New York Post Corporation.

367. Philosopher Sidney Hook:

A philosopher in his own life need be no more wise than a physician needs to be healthy.

368. Advertisement for skin lotion:

You've seen land crack and dry when it loses its essential moisture; the same thing can happen to your skin when it loses its moisture.

*369. Overheard:

If we find it necessary to tip waiters and other hotel servants, why should we not similarly reward the bus driver, the clerk, or the doctor? Either they should be included or hotel tipping should be abolished.

370. Old analogy:

The universe as a whole is like a watch in that it exhibits mechanical order. But watches cannot come into existence without a creator who is responsible for the order they exhibit. Therefore, the universe could not have come into existence without a creator who is responsible for the order it exhibits.

371. From Justice Holmes:

We have seen more than once that the public welfare may call upon the best citizens for their lives. It would be strange if it could not call upon those who already sap the strength of the State for these lesser sacrifices, often not felt to be such by those concerned, in order to prevent our being swamped with incompetence. It is better for all the world, if instead of waiting to execute degenerate offspring for crime, or let them starve for their imbecility, society can prevent those who are manifestly unfit from continuing their kind. The principle that sustains compulsory vaccination is broad enough to cover cutting of the Fallopian tubes.

372. From B. F. Skinner:

The concept of responsibility is particularly weak when behavior is traced to genetic determiners. We may admire beauty, grace, and sensitivity, but we do not blame a person because he is ugly, spastic, or color blind. Less conspicuous forms of genetic endowment nevertheless cause trouble. Individuals presumably differ, as species differ, in the extent to which they respond aggressively or are reinforced when they effect aggressive damage, or in the extent to which they engage in sexual behavior or are affected by sexual reinforcement. Are they, therefore, equally responsible for controlling their aggressive or sexual behavior, and is it fair to punish them to the same extent? If we do not punish a person for a club foot, should we punish him for being quick to anger or highly susceptible to sexual reinforcement? The issue has recently been raised by the possibility that many criminals show an anomaly in their chromosomes. The concept of responsibility offers little help. The issue is controllability. We cannot change genetic defects by punishment; we can work only through genetic measures which operate on a much longer time scale. What must be changed is not the responsibility of autonomous man but the conditions, environmental or genetic, of which a person's behavior is a function.

B. F. Skinner, *Beyond Freedom and Dignity* (New York: Knopf, 1971), p. 75.

373. From Ayn Rand:

By what conceivable standard can the policy of price-fixing be a crime, when practiced by businessmen, but a public benefit, when practiced by the government? There are many industries in peacetime—trucking, for instance—whose prices are fixed by the government. If price-fixing is harmful to competition, to industry, to production, to consumers, to the whole economy, and to the "public interest"—as the advocates of the antitrust laws have claimed—then how can that same harmful policy become beneficial in the hands of the government? Since there is no rational answer to this question, I suggest that you question the economic knowledge, the purpose and the motives of the champions of antitrust.

Ayn Rand, *Capitalism: The Unknown Ideal* (New York: NAL, 1966), p. 52.

False Cause

False cause refers to an argument that suggests events are causally connected when in fact no such causal connection has been established. The fallacy lies at the root of most superstitions (which are mixed-up notions of what causes what), and the list that follows begins with a sampling of these forms of the fallacy (exercises 374 to 382).

Aside from simple superstition, a typical form of the fallacy consists of assuming that because two events occur at roughly the same time or one occurs immediately prior to the other, the one caused the other. Exercises 383 to 395 are examples of the fallacy in this form. Here we have to remind ourselves that sequence alone is no proof of consequence or to be careful not to confuse cause with coincidence.

If immediate succession is not necessarily proof of causation, much less so is somewhat remote succession. Exercises 396 to 399 are a few examples of the fallacy in this form.

Another typical form of the fallacy is to reverse the cause-effect relationship entirely (exercises 400 to 404) or to forget that two things may be related not in the sense of one being the cause of the other but because both are effects of a common cause (as in exercises 405 to 407).

As in other cases, a few humorous examples of the exploitation of this fallacy (exercises 408 to 413) have been added here.

374. From Shakespeare's *King Lear* act 1, scene 2:

Earl of Gloucester: These late eclipses in the sun and moon portend no good to us: though the wisdom of nature can reason it thus and thus, yet nature finds itself scourged by the sequent effects: love cools, friendship falls off, brothers divide: in cities, mutinies, in countries, discord; in palaces, treason; and the bond crackt 'twixt son and father.

375. In 1577, during one of the plagues, Thomas White preached a sermon at Paul's Cross that contained the following remarks:

Looke but uppon the common playes of London, and see the multitude that flocketh to them and followeth them: beholde the sumptuous Theater houses, a continuall monument of London prodigalitie and folly. But I understande they are now forbidden bycause of the plague. I like the pollicye well if it hold still, for a disease is but bodged or patched up that is not cured in the cause, and the cause of plagues is sinne, if you looke to it well: and the cause of sinne are playes: therefore the cause of plagues are playes.

Quoted in the preface of Hardin Craig, *Shakespeare: The Complete Works* (Glenview, Ill.: Scott, Foresman, 1961), p. 27.

376. Conversation between King Charles II and Milton:

"Think you not," said Charles II to Milton, after the poet had become blind, "that the crime which you committed against my father must have been very great, seeing that Heaven has seen fit to punish it by such a severe loss as that which you have sustained?" "Nay, sire," Milton replied, "if my crime on that account be adjusted great, how much greater must have been the criminality of your father, seeing that I have only lost my eyes, but he his head!"

*377. Chain letter:

TRUST IN THE LORD WITH ALL THINE HEART AND
ACKNOWLEDGE HIM AND HE WILL SUSTAIN THEE,
AND LIGHT THY WAY

This prayer has been sent to you for good luck. The original is from the Netherlands. It has been around the world nine times. The luck has been brought to you. You are to receive good luck within four days of receiving this letter. This is no joke. You will receive it in the mail.

Send copies of this letter to people you think need good luck. Do not send money. Do not keep this letter. It must leave you within ninety-six hours after you receive it. An RAF officer received $70,000 but lost it because he broke the chain. While in the Philippines, General Walsh lost his life six days after he received this letter. He failed to circulate the prayer. However, before his death, he received $775,000. Please send twenty copies and see what happens to you on the fourth day.

This chain comes from Venezuela and was written by St. Anthony de Cadif, a missionary from South America.

Since this chain must make a tour of the world, you must make twenty copies, identical to this one, and send them to your friends, relatives and acquaintances. After a few days, you will get a surprise. This is true, even if you are not superstitious.

Take note of the following: Constatine Diaz received the chain in 1953. He asked his secretary to make twenty copies and send them. A few days later, he won a lottery of two million dollars in his country. Carlo Cravit, an office employee, received the chain and forgot it and in a few days, he lost his job.

He found the chain and sent it to twenty people. Five days later he got an even better job.

Darin Wirchild received the chain and not believing in it, threw it away. Nine days later he died.

For no reason whatsoever should this chain be broken.

378. Popular belief:

Since 1840, all the United States presidents who have been elected in even-numbered years in multiples of twenty have died in office: Harrison, 1840; Lincoln, 1860; Garfield, 1880; McKinley, 1900; Harding, 1920; Roosevelt, 1940; and Kennedy, 1960. The president elected in 1980 will suffer the same fate.

379. News report from the Associated Press:

LOUISVILLE, KY. (AP)—The Rev. Riner has crusaded against many things he considers scourges upon the Louisville community. He has loudly denounced area productions of the rock musicals "Hair" and "Jesus Christ Superstar." And after a tornado ripped through sections of Louisville in 1973, he marched with his "Jesus Club," warning passersby to mend their evil ways or expect another devastating storm.

Los Angeles Herald-Examiner, December 13, 1975.

380. News report:

The U.S. gave Japan baseball and now Japan is giving baseball Nichiren Shoshu, a turned-on version of Buddhism and 20th century power of Positive Thinking. Nichiren Shoshu claims 200,000 members in the U.S., including Los Angeles Dodger outfielder Willie Davis, 32. "This religion is simply a must for sportsmen," said Davis while on pilgrimage to the head temple at the foot of Mount Fuji. "I was never a great home-run hitter. I hit only ten home runs in the 1971 season. Last year I suddenly ended up hitting 19 because I chanted my prayers every morning and night and before every game." Why did his batting average drop from .309 to .289? "I didn't pray hard enough."

381. Letter to the editor:

With Tom Dempsey so long gone, why don't the Rams throw away No. 11 and make Pat Haden No. 10 again? In all his years as No. 10, the quarterback never looked like he did as No. 11 against Minnesota. As No. 10 doesn't Haden always go to the Rose Bowl?

382. **From Thomas Harris:**

Parent: Illegitimate, you know.
Parent: Oh, *that* explains it.

Thomas A. Harris, *I'm OK—You're OK* (New York: Avon Books, 1973), p. 97.

383. **From a vocabulary book:**

Your boss has a bigger vocabulary than you have. That's one good reason why he's your boss.

W. Funk & L. Lewis, *Thirty Days to a More Powerful Vocabulary* (New York: Funk and Wagnalls, 1970), p. 3.

*384. **Letter to the editor:**

More and more the government is originating welfare legislation designed to alleviate our citizens from slavery of economic necessity. At the same time, however, we are seeing an increase in immorality, alcoholism, and suicide among our citizens. This makes it clear that to get rid of this upsurge in immorality, alcoholism, and suicide, we must stop the government from instituting welfare measures.

385. **News report from the Associated Press:**

Fuel Crisis Tied to Drinking Rise

EVANSTON, ILL. (AP)—The president of the Women's Christian Temperance Union said Sunday that people were turning to drink to escape the worries of the troubled national economy. "Liquor dealers admit that since the energy crisis began, the consumption of alcoholic beverages has greatly increased," said Mrs. Fred Tooze, head of the national anti-alcohol group. Mrs. Tooze said the need to conserve gasoline would cause people to stay home and drink more, creating broken homes and harming the mental capacity of the nation's work force.

386. **From a column by Mike Royko:**

Damning rock music for its "appeal to the flesh," a Baptist church has begun a campaign to put the torch to records by Elton John, the Rolling Stones, and other rock stars. Some $2,200 worth of records were tossed into a bonfire this week after church officials labelled the music immoral. The Rev. Charles Boykin, associate pastor and youth director at the Lakewood Baptist Church of Tallahassee, Florida, said he had seen statistics which showed that "of 1,000 girls who became pregnant out of wedlock, 986 committed fornication while rock music was being played."

"Watch Out for That Old Devil, Syncopation," *Los Angeles Times,* December 7, 1975.

387. Advertisement for sound equipment:

Don't ever let anyone tell you that stars aren't made. It happens everyday. Professional sound equipment can make it happen a lot sooner.

388. Advertisement:

"My hair isn't all that's changed."

Brighten up your spirits. Give your hair a Frost & Tip look. It's easy. You can do anything from a single streak to a full head of frosting.

And remember, you won't be changing your whole haircolor, you'll just be adding excitement.

With Clairol's Frost & Tip, you can do beautiful new things for your hair. And you. And that's why it's the number one frosting kit.

Frost & Tip.
It's more than a look. It's a feeling.

*389. Overheard:

That cream cleared up her complexion within six months.

*390. Overheard:

Have you noticed how the sales went up after we instituted our new advertising campaign? Our success is obvious.

391. Letter to the editor:

In some countries where atheism has spread in recent years the suicide rate has also gone up. Therefore loss of belief in God is a cause of suicide.

392. Overheard:

The city never had this problem with garbage disposal until *he* became mayor.

393. Overheard:

No sooner did they start to fluoridate the water but my friends began dying of heart disease. It just doesn't pay to tamper with nature!

394. Overheard:

I was so sick after I ate those green apples that I swore I would never eat another apple again.

395. Overheard:

Janice got a typewriter for her birthday and now she gets As in everything. If I had a typewriter....

*396. From the "Better Half" by Barnes:

"I knew some day I'd regret letting all those gorgeous women run their fingers through my hair."

397. Letter to the editor:

"Tis a headline writer who can claim the Carter victory. Ford won New York state but lost New York City, and it was because of the headline "Ford to N.Y.: Drop Dead" that the city was lost. For want of the city the state was lost; for want of the state the electoral college, and so the presidency, was lost. N.Y. to Ford: Drop Dead.

398. From David Reuben:

Adolf Hitler slipped three cents to a Viennese whore in 1910 (that was the going rate for street girls then) and she slipped him a few million bacilli of syphilis. Ironically, that one case of syphilis besmirched a cultured and civilized nation with depravity and sadism previously unknown in the history of the world.

David Reuben, *How to Get More Out of Sex.* New York: David McKay Co., 1975, p. 222. Copyright © 1974 by David Reuben. Reprinted by permission of David McKay Co. and Harold Matson Company, Inc.

*399. From *The Only Diet That Works:*

Do you like being served a plate piled high with food? Or does the overgenerous plate make you lose your appetite? If the latter, you may be still rebelling, many years after, against the parent who lovingly insisted you overeat. If the former, that well-heaped plate may simply be a heart-warming (and unconscious) reminder of mother's tender loving care and family table generosity.

Do you like midnight snacks? It may simply be that you need the maternal reassurance you once had nightly as a child. Or do you have occasional "uncontrollable" desires for a sweet? This can be a simple matter of sugar-need, but it can also be a matter of feeling unappreciated, and wanting the sweet that was your ten-year-old reward for being good.

Do you eat a limited menu?

Are you afraid of new foods? Yes? Then, when you were a child, were you by any chance frightened by and consequently made to be afraid of, strangers? There can be a connection. And there is considerable evidence incidentally that well-balanced people tend to have fewer food aversions than the maladjusted.

Excerpt "Do you like . . . than the maladjusted" page 38 of *The Only Diet That Works* by Herbert Brean, Copyright © 1965 by Herbert Brean. By permission of William Morrow and Company, and Harold Matson Company, Inc.

*400. From early Greek physics:

Night is the cause of the extinction of the sun, for as evening comes on, the shadows arise from the valleys and blot out the sunlight.

*401. Editorial:

If strong law enforcement really prevented crime, then those areas where police patrols are most frequent would be the safest and the best protected. Actually, the very reverse is true, for in such areas even one's life is in danger, and crimes of all kinds are more common than in other areas where police patrols are infrequent.

402. Sitcom:

Tortoises live for over a hundred years; therefore if you move slowly, you will live long.

403. Overheard:

The greater the number of laws and enactments, the more thieves and robbers there will be.

404. A woman spent most of a Saturday afternoon preparing a gourmet dinner for a boyfriend. He ate it with obvious pleasure, then said:

That seemed like one of the best meals I've ever eaten. My body chemistry must have been just right.

*405. Overheard:

In the years prior to the outbreak of recent wars there has always occurred an increase of armaments among the belligerents. It is obvious, therefore, that increased armaments is the cause of war.

*406. Letter to the editor:

If religion prevented crime, the cities, where churches abound, would show a lower crime rate than the ten thousand rural communities said to be without churches. Actually the city rate is higher.

407. Editorial:

On every occasion in which major tests of atomic bombs have been made, serious storms have been reported in various parts of the United States. It is evident that these tests must be stopped unless we wish to change the world's weather pattern.

408. Sitcom:

Last week I got into trouble through imbibing too much brandy and gin. The other day it was ale and gin. And I remember that two months ago I spent a sorry day after an evening of beer and gin. I see, accordingly, that it is the gin that must be responsible. I must give that up and I shall be all right.

409. Historical note:

While General Grant was winning battles in the West, President Lincoln received many complaints about Grant's being a drunkard. When a delegation told him one day that Grant was hopelessly addicted to whiskey, the president is said to have replied, "I wish General Grant would send a barrel of his whiskey to each of my other generals."

410. From the pen of the great Yiddish storyteller and humorist Sholem Aleichem:

The young man standing next to him asked, "What time is it?" The old man refused to reply. The young man moved on. The old man's friend, sensing something was wrong, asked, "Why were you so discourteous to the young man asking for the time?" The old man answered, "If I had given him the time of day, next he would want to know where I'm going. Then we might talk about our interests. If we did that, he might invite himself to my house for dinner. If he did, he would meet my lovely daughter. If he met her, they would fall in love. I don't want my daughter marrying someone who can't afford a watch."

*411.

Los Angeles Times, May 3, 1978, Part 4. Copyright 1978 Universial Press Syndicate. Reprinted by permission.

412.   From *Tennis* magazine:

Have you ever noticed that there's an interesting correlation between the strength of feminism and the length of skirts? Traditionally, as women stand up to be counted, their hemlines rise, too.

"Are Tennis Skirts Too Short?" Reprinted with permission of *Tennis* magazine, May 1976, Copyright © Tennis Features, Inc., p. 88.

413.   After hearing Jascha Heifetz perform in London at the age of nineteen, George Bernard Shaw wrote him:

My Dear Heifetz:

Your recital has filled me and my wife with anxiety. If you provoke a jealous God by playing with such superhuman perfection, you will die young. I earnestly advise you to play something badly every night before going to bed, instead of saying your prayers. No mortal should presume to play so faultlessly.

G. Bernard Shaw

Irrelevant Thesis

In the case of the fallacy of *irrelevant thesis* an attempt is made to prove a thesis or a conclusion that is not the one at issue or, alternately, to refute a thesis different from the one at issue in the other person's argument.

Like many of the other fallacies, it goes under a variety of names, including *irrelevant conclusion, ignoring the issue, befogging the issue, diversion,* and *red herring.*

Diverting attention away from what is at issue can take many forms: in addition to plain irrelevance (as exemplified in exercises 414 to 419), it may take the form of gross exaggeration (exercises 420 to 426), humor (exercises 427 to 431), or anger (exercises 432–433).

*414. Henry Kissinger, talking about the 1972 war between India and Pakistan over Bangladesh:

There have been some comments that the administration is anti-Indian. This is totally inaccurate. India is a great country.

415. Mother:

Steven, I don't care if you don't like spinach. Don't you know there are people starving all over the world.

416. Recommendation:

I am sure that Bill is best qualified for the engineering scholarship. He worked hard for his high marks, unlike some of the applicants, for whom good grades came easily. Bill is athletic and friendly. And furthermore, both his father and grandfather before him won this same scholarship.

*417. Overheard:

It is silly to say that modern killer-diller comic books hurt youngsters; they are just the modern version of the dime novel.

418. Overheard:

Jones: We've got a lot to learn from the USSR.
Smith: D'ya mean to say we've got nothing to teach them?

419. News report:

If you've joined the gay boycott of Florida orange products, you've made a healthy move. It turns out that Florida oranges are dyed with Citrus Red 2, a carcinogen, while California oranges are not.

"Gay Boycott Stops Cancer," *Berkeley Barb,* January 27, 1978.

420.

USC Daily Trojan, May 15, 1978.

*421. Talk show:

> Guest: I maintain that the government should increase its welfare benefits.
>
> Host: Ladies and gentlemen, Mr. Jones would recommend that the country go Communist!

422. Politician:

Ladies and gentlemen, my opponent has stated that he believed in law and order and that if elected he would do everything possible to make the streets safe so our wives could walk the streets at night. What, I ask you, does he want to do, make hookers of our wives?

*423. Letter to the editor:

Vegetarianism is an injurious and unhealthy practice. For if all people were vegetarians, the economy would be seriously affected and many people would be thrown out of work.

*424. Overheard:

I regret I find myself unable to support this no doubt deserving cause. It is quite impossible for me to respond to all the charitable appeals that are made to me.

425. Overheard:

No, I'm not going to buy an electric jigsaw. If I did, you would want a circular saw, and then a power drill, and soon you would want a whole workshop. We can't afford all that.

*426. Student essay:

As I understand it, communism proposes that we take all the money and divide it equally among everyone. This is silly. In no time at all, the smart ones would have the biggest share again.

427. Sitcom:

Teacher: How much is three times five?
Pupil: I hate math!

428. Sitcom:

Woman: (too fat to wear the outfit):
By the way, what do you think of my new outfit?
Friend: It's certainly in style.

429. When O.J. was playing for the Trojans:

Reporter: Why does O.J. carry the ball so often?
Coach Mackay: Why not? It isn't very heavy.

*430. From James Boswell:

We talked of the education of children; and I asked him what he thought was best to teach them first. JOHNSON. "Sir, it is no matter what you teach them first, any more than what leg you shall put into your breeches first. Sir, you may stand disputing which is best to put in first, but in the meantime your breech is bare. Sir, while you are considering which of two things you should teach your child first, another boy has learned them both."

James Boswell, *The Life of Samuel Johnson* (1791), notes and introduction by Herbert Askwith (New York: Random House, n.d.), p. 273.

431. Conversation between two flower children:

She: Do you love me?
He: I love everybody.

432. From Eric Berne:

Little girl: Mommy, do you love me?
Mother: What is love?

Eric Berne, *Games People Play: The Psychology of Human Relationships* (New York: Grove Press, 1976), p. 94.

*433. George Meany, AFL president:

To these people who constantly say you have got to listen to these younger people, they have got something to say, I just don't buy that at all. They smoke more pot than we do and if the younger generation are the hundred thousand kids that lay around a field up in Woodstock, N.Y., I am not going to trust the destiny of the country to that group.

Quoted in the *New York Times,* August 31, 1970. ©1970 by The New York Times Company. Reprinted by permission.

ANSWERS TO EXERCISES
PART TWO

Sweeping Generalization

146. This and the next few examples provide useful practice in isolating generalizations, for they can take many different forms.

147. "Those who thrust a knife into another person" does not look like an ordinary generalization ("All...") but serves as such here.

148. This one has more of the appearance of a typical generalization ("Everyone...").

150. Running a government is much too complex for one person to be able to do it alone.

151. Narcotics are habit forming but the circumstances here are special: the drug will apparently be administered by one with the proper knowledge and under guarded conditions and for probably a limited period of time. It is therefore not likely to lead to the same results.

153. Blue-collar labor, although indispensable, is not the only, or even the major kind of labor that creates the wealth of a country.

Hasty Generalization

158. There are, of course, good reasons the vehicle in question is permitted to do what it does. As an exceptional case it does not support the conclusion drawn.

162. Newspapers generally do not create the news but only report it. Evidence from more than one situation would be needed to support the conclusion in question.

163. Some people have indeed accomplished a great deal without the benefit of formal schooling, but they are hardly in the majority and hardly representative of the average person.

164. What may work very well in Alaska may not work so well in places where the conditions and climate require something rather different in the way of "toughness."

165. Absurd. Science has more than earned our confidence in having accomplished what it has. No doubt in time it will succeed in explaining the origin of life, too.

167. Two mixed-up orders are hardly sufficient to condemn every clerk in the store. In addition note when this occurred—the Christmas rush season.

168. One meeting with one group of students hardly warrants this negative response.

171. The parents at Berkeley may not be representative of parents elsewhere. A larger survey is needed to determine this.

172. If Burgess has not made a wider study of modern education and his young son's experiences at that one school is the only basis he has for making the statements in question, then Burgess is hardly justified in making them.

173. Many non-Pisceans display the same characteristics, and no doubt many Pisceans do not. A much wider sample would be needed to establish the link in question.

174. A great nose is not necessarily a sign of greatness. Some great human beings have been cursed with rather large noses and so have a great many ordinary folks.

175. It takes more than a certain look in the eyes to establish insanity. As in the previous case, there is no necessary or essential connection between the two.

176. Some manufacturers spend more money on the package than on what is put in it, knowing how gullible we are. Since the bottle does not smell, it does not matter how elegant it is; it will or can have no effect on what is inside it.

179. Although being fond of children is an important factor here, it alone could not guarantee the person would make a fine kindergarten teacher. Intelligence, dedication, patience, perseverance—these may be even more important qualities to look for.

180. The fact that Heyerdahl's raft drifted and came to rest on the island of Barbados does not by itself establish that Barbados was the first landing place of the ancient Egyptian voyagers. The winds and currents may very well have carried them elsewhere (assuming, of course, they attempted the journey in the first place).

181. This is an example of another form the fallacy of hasty generalization can take—basing a conclusion (hastily again) on one-sided evidence. It may be true that state-owned industries encourage featherbedding and absenteeism, but to abandon them entirely is like throwing the baby out with the bath water, getting rid of the good with the bad. Why not try instead to correct the abuses?

183. What is stated is no doubt true, but it is only part of the story. What is neglected is some recognition of the need and value of international trade. Other nations are able to do some things better and more economically than we can and it may make better sense to let them produce these things and sell them in our country in exchange for things we have that they need or that we can produce more economically than they.

184. Times change. What has worked well in the past may not work well now. There are more of us now; the world is a more complex place now than it was formerly, and our needs are more varied and complex, too. Their satisfaction and world peace cannot be left completely to chance.

Bifurcation

189. This is probably the best-known commercial and one of the most catchy. What is objectionable about such appeals is that they try to cut off our critical thinking about a product by channeling our view of it into an either/or polarity that suits their aim. It is comforting to realize, as the following news report from *Time* Magazine shows, that something is finally being done to challenge the questionable tactics advertisers resort to.

In promotions stretching back to 1921, Warner-Lambert has asserted that its Listerine mouthwash helps prevent colds and sore throats. Last week that claim was finally snuffed out by a fatal regulatory infection called truth in advertising. The Supreme Court declined to review a lower court decision upholding a 1975 Federal Trade Commission order: the company must not only stop making the claim but specifically advertise that it is not true. In its next $10 million worth of Listerine ads—about a year's budget—Warner-Lambert must insert this statement: "Listerine will not help prevent colds or sore throats or lessen their severity." In the course of its review, which began in 1972, the FTC found that Listerine was no more effective in combatting colds than warm water. Doubtless Warner-Lambert will bury the admission as inconspicuously as possible in ads declaring that Listerine does cure bad breath—another old claim.

Nonetheless, last week's Supreme Court refusal to review the order is a significant boost for the FTC. The agency in the past seven years has forced other companies to run "corrective" ads asserting in effect that their previous ads made false claims. Companies bowing to such orders include ITT; Continental Baking for Profile bread (whose claimed fewer calories per slice, the FTC charged, was attained simply by making its slices thinner); Ocean Spray for cranberry juice; and Amstar for Domino sugar. All signed consent decrees; Warner-Lambert was the first to ask the courts to rule that it did not have to take back its previous claims. Now that it has definitively lost, says a jubilant FTC staffer, "I think we will see more corrective ads in the future."

April 17, 1978, p. 83. Reprinted by permission from *Time, The Weekly Newsmagazine;* Copyright Time Inc. 1978.

191.   There are obviously other artificial stimulants; nor are artificial stimulants the only way to keep awake—a cold shower, fresh air, a good night's sleep, work, too.

199.   What is objectionable about this and many other clichés based on the fallacy is that they force us into a position of choosing between two alternatives when in fact our choices are not so limited.

201.   Famous people are not immune to the fallacy. Our reply here could be that we might try to change our circumstances, or, failing that, ourselves

202.   Not always. Sometimes we become greatly diminished as a result of inordinate pain and suffering and injustice.

203.   It may sometimes be more courageous to try to survive and fight another day.

210.   Professors need to be challenged and inspired too, and bright students help. Too many overly dull students can prove discouraging.

211.   What is overlooked here is the fact that large numbers of students are neither overly conscientious nor hopelessly lazy, and for them "impending examinations" are neither unnecessary nor useless.

213.   Why must it be either the one or the other? Why not a combination of both approaches?

214.   Again, an example of highly polarized thinking. Who says it must be one or the other? We may have to settle for something less than absolute freedom in order to avoid succumbing to communism.

217.   Are these indeed the only possibilities? Safeguards to protect our natural resources can be found; and we are not necessarily bound to deal only with Arab states. In addition, new reserves keep being discovered and the world supply appears to be greater than first believed.

219.   Is it not possible that a safe amount of insecticides could be used so that widespread hunger would not occur? Limiting the use of insecticides to either too much or none at all makes our situation appear more desperate than it is.

222.   Since Christians represent a minority of the world's population, the rest are therefore condemned by the argument to a life of torment in hell.

223.   This is a moving and eloquent passage but flawed very much in the same manner as the others are: the world is neither wholly a prison nor wholly a madhouse, nor are we either wholly egoistic or wholly altruistic. The language is impressive and beautiful but not that significant.

Begging the Question

225.   This is the typical form of this argument in this context: the Bible is true because the Bible itself says so. But, of course, that is the central question: how do we know that what the Bible says is really so?

226.   This is a little more verbose and a bit more confusing but essentially the same argument.

227. College catalogs, as every student sooner or later discovers, often are full of inaccuracies and not infrequently contain a good deal of misinformation.

230. Since "cannot happen" is simply another way of saying "impossible," the "argument" amounts to saying "miracles are impossible" because "they are impossible." The person may be right about this, but what is offered is not an argument.

231. Since "school" is essentially the same as "book learning" and "not worthwhile" the same as "does not pay off," the second part of the statement (the premise) simply repeats the first part (the conclusion), with the result that nothing is confirmed.

234. Combustible means being able to burn and therefore offers no evidence for the conclusion (namely, *why* hydrogen burns). To say it burns because it burns is to say nothing.

236. If people do not have the ability to "distinguish what is right and what is wrong," then they lack "moral values," for having moral values means being able to distinguish between what is right and wrong. The "argument" therefore does not tell us why modern moral values are deteriorating; it merely restates the point. Had the person argued that people's values were deteriorating because they have lost their belief in God or their self-respect, that would constitute an attempt to provide a reason for believing in a certain conclusion and not merely a matter, as here, of repeating it.

239. A lower price policy would mean the store would naturally have to reduce its prices. It is silly to say that the low prices are a result of "our lowered price policy." What is the reason behind the price reduction? Is the store trying to undercut a nearby store? Has business been so poor that the store needs to attract new customers?

242. "Free" and "liberty," "all" and "universal," and "people" and "humanity," are identical in meaning; the "argument" therefore merely uses different words to say the same thing, with no reason given in support of the contention that all people should be free.

243. Since "those who betray our country" is another way of saying "traitor" and "right" is the same as "properly justified," the premises of the "argument" simply repeat what has been asserted in the conclusion. We have here therefore only an assertion and not an argument.

246.   Since "indiscriminate" means lacking the ability to "distinguish between deserving and undeserving cases" and "reprehensible" means essentially the same as "wrong," we once again have an "argument" in which the words could be interchanged without making any difference to what is conveyed.

250.   "Each individual should enjoy a liberty perfectly unlimited, of expressing his sentiments" is simply another way of expressing "every man" should be allowed "an unbounded freedom of speech" and the phrase "highly conducive to the interests of the community" is another way of saying "advantageous to the State," with the result that nothing new is presented in support of the original claim.

251.   How could a person live like a millionaire without being one? Of course one could not. This is not, strictly speaking, a fallacy but the use of the same device to express wit—in this case just an amusing or engaging way of saying "I would like to be a millionaire."

253.   If this is not an attempt to be funny, then it is simply absurd, for obviously if you knew exactly when the snake was born, you would know its age. The question is, is it possible to tell how old a snake is without knowing when it was born?

254.   If the husband knew where he had left his cuff links, he would not have needed to ask his wife where they were! On the psychological level such an answer is "unhelpful"; on the logical level such an answer is simply "meaningless."

257.   If he does not want his wife to speak to anyone, then he has a pretty severe case of it—for that is what we mean when we say a person is jealous.

259.   Once again we have a case of circular reasoning in operation: "liberals don't like nuclear plants" because "it is their philosophy." And what is their philosophy? Well, "not to like nuclear plants" (which explains nothing about why, in particular, liberals feel this way). A is so because of B; and B is so because of A.

260. The fan does not answer his friend's question: What makes the Broncos a better team, one able to win more games than the other team? They are better (A), he says, because they won more games (B); and they won more games (B) because they are a better team (A).

263. If we could be sure he indeed does talk with angels, then we could accept his word for it. But does he really talk with them? That question is begged. In other words, no one "capable of talking with angels" would lie; but first we need to find out whether he does talk with them.

266. Whether there might be something "uncaused" or "self-caused" is the whole issue here. To dismiss it out-of-hand as done here is to fail to recognize what the whole question is about. The "argument" simply assumes that everything that happens has a cause other than itself.

270. The first part of the "argument," which states that "reality in itself must be as it appears to the five senses," implies that is the only way we can know reality. But that is precisely the question at issue here and must not be begged.

272. This is an interesting example, made even more interesting when we find Webster's dictionary giving the following definition of clock: "an instrument for the measurement of time by the motion of its parts."

275. Cause/effect being like husband/wife correlative terms, you cannot have one without the other. The real question here is therefore whether uncaused "events" (not "effects") are possible. In other words, or course if "causes" did not exist we would have "effects without causes," for it is inherent in the ideas of "cause" that there should be "effects." But this does not face the question whether there might be uncaused "events."

276. As Professor Pospesel points out here, Spurney simply presupposes just what he is supposed to prove, that a human soul exists.

Question-Begging Epithets

277. Words like "deliberate" and "plot" and "against" carry with them implications that require proof and support and should not be allowed to go unchallenged. If a few people get together to discuss the political situation, this hardly is a matter of engaging in a "plot"; and even if their discussion turns highly critical of the present administration, this does not mean they are disloyal. Our tradition of a loyal opposition should be a help here in resisting the innuendos.

279. "Measure," "calculated," and "subvert" are all question-begging dyslogistic terms; "just" and "aspirations" are question-begging also, but of a eulogistic or complimentary sort. Both are objectionable because they assume the attitudes they try to generate are justified when no evidence for this has been offered.

280. This example is similar to the previous one in using both dyslogistic and eulogistic terms: calling the proposal "offensive" is an attempt to condemn it out-of-hand; and saying that people of an "enlightened conscience" will certainly find it so is to discourage even the thought of challenging its truth—for who would like to regard himself or herself as possessing an unenlightened conscience?

282. The remark tells us a great deal about Milton Gross and not too much about Kunen. Instead of "heavily haired," he might have written "who wears his hair long" (and avoided the implication of someone "messy" and "dirty"); and the same applies to such terms as "wealthy" and "post adolescent."

283. Of all the question-begging terms used in this statement ("degrading," "dole," "loafers," "parasites"), none is as offensive or as contemptuous as the word "them." Why?

284. Using carefully chosen key words, the author of this letter builds up the Throneberry family while tearing down the union. The Throneberrys are "hard-working," "honest," and "law-abiding"; the union people are "goons" and "militants" who engage in "physical violence" and "the destruction of property." We are not given a word as to why there was a strike (one which, for all we know, may have been justified).

285. This brief editorial is loaded with question-begging epithets: "destructive raids," "perpetrated," "rabid," "bust," "menace," "Gestapo tactics," and so on. No doubt some government agents on some occasions have tackled their assignments with

more zeal than may have been called for, but this hardly justifies ranking or comparing them with the Gestapo or those otherwise deranged ("rabid").

287. This letter is filled with insulting question-begging epithets, the worst being "our little vacationing soldier boys" (implying they are just having fun and are not involved in any serious missions). And where was the letter writer when he observed this behavior? At the barracks or at the local bar? And how many did he see? How many could he possibly have seen? And what was *he* doing when he made these observations? There are many unanswered questions here.

289. To comment on the first paragraph: had Senator Byrd merely been interested in informing the public at that time about certain facts (and not simply trying to alarm the public and venting his anger), he might have used such terms as "change," "alter," or even "influence" instead of "distort." He would not have used the phrase "advocate the overthrow of our system of government," which is the quickest way we have devised to raise the ogre of communism; nor would he use such customarily explosive and manipulative terms as "corrupt and pervert" or "purged." But Senator Byrd was at that time probably less interested in presenting certain facts to the public than in raising an emotional storm. A final point: had a survey been conducted of these so-called "fuzzy-minded, phony liberals" to see who their "heroes" were, one wonders how often the names of Che Guevara and Fidel Castro would in fact have come up.

Complex Question

293. The question assumes the new building was supposed to be given to the school when this may not have been so.

297. We might reply yes, this policy is designed to lead to deflation but not a ruinous one.

301. This is a case of combining question-begging epithets and complex question. Is going to school a "waste" of time? What constitutes a "man's work" or a "contribution"? Many business firms today give the executive positions to those with college degrees, believing those with a college education will probably make a greater contribution to society than those without such an education.

304. The person is assuming that all high schools have "full-to-the-gills student parking lots" and that every teenager feels a car is his or her birthright. How many surveys has the speaker conducted to confirm this?

306. The interviewer unfairly assumes here that the senator finds it hard to come to a conclusion. Why not regard the senator's hesitance as expressive of his cautious approach to difficult and complex problems?

312. This argument assumes there was a time before time began. How could there have been a time when there was no time?

315. This is an enormously clever advertisement. Yes, kids eat a lot of sweets and yet stay thin. They lead an active life and work off the calories. Adults who lead a much more sedate existence can expect the intake of sweets to have visible results. The claim that sugar is not fattening because kids eat a lot of it and are not fat is at least devious.

316. This advertisement tries to make you believe their mascara will make your lashes beautiful once again when it may not improve them at all.

321. The sales representative is trying to help the customer make up his or her mind by assuming the customer has already decided to buy the item—the only question remaining being the type of contract preferred.

325. They are assuming here that the fountain pen will last a lifetime when many things could go wrong with it before long.

327. Any overseas flight will wear one out because of the time change alone. The suggestion that they will be able to entertain and relax you enough so you will not be worn out is questionable.

329. No family is ever going to be that secure with the husband gone, and very few people would want to set aside a large enough portion to provide for such security. The family might be much better off if the man's wife were trained for a well-paying job or position.

330. The husband in this interchange assumes he cannot find the can opener because his wife "hid" it from him. His wife cleverly decides to diffuse a potential blowup by admitting she "hid it next to the tablespoons."

Special Pleading

331. Dr. Szasz's point is that both probably have the same personality traits or exhibit the same behavior, but the patient is labeled "narcissistic," "inhibited," and so forth, whereas the analyst is said to have "self-esteem" and "self-control."

336. What was this person doing snooping in his or her roomer's desk? The speaker is therefore just as untrustworthy as the person condemned.

338. Making someone an offer they cannot refuse (or as here, "just trying to convince them...") is what we mean by the term "blackmail."

345. The farmer unfortunately cannot see that he and his friends have just as much imagination as the blacks do. The whites want to see what they want to see (and do), and the same is true of the blacks. But so blinded are we that often, like that farmer, we do not recognize we are resorting to a double standard. In this case, in addition, both types of viewers are suffering from an overheated imagination. This is a lovely (if such things may be called "lovely") example of the fallacy at work.

347. The Israelis are supported by "imperialist" funds and technology; but the Arabs, who at that time were given financial, political, and military aid by the Soviet Union, probably the world's largest and most ruthless colonial power, are presumably nonimperialists. Further, the "argument" tries to persuade us that the Palestinians, but not the Israelis, have a right to self-determination.

348. This kind of unfortunate rhetoric is unlikely to lead to peaceful solutions. Apparently we are asked to believe that the Jews who struggle for national liberation are racist tools of imperialism, while Arabs who want to "purge the Zionist presence" are merely fulfilling their national destiny. Furthermore, Jews who were born and educated in Israel of parents who were also born and educated there are not Palestinians and have no right to be there, but Arabs born in Kuwait to fathers of Palestinian ancestry who were born in Egypt are Palestinians and have a right to "return" to their "homeland." However, a Jew whose grand-father left Egypt for the land of Israel in 1920 is an Egyptian!

False Analogy

350. This is not a very meaningful comparison. When three hundred people die from choking accidents, only they die; but in the case of nuclear power plant accidents, not only countless numbers of people might suffer from the disasters, but the environment might be adversely affected for thousands of years as well. Furthermore, just because there are more choking accidents than nuclear power plant accidents in a year does not mean one may not occur. Chances are the more nuclear power plants are built, the more likely it is an accident will eventually occur.

352. It is true that a razor will get blunt from constant use, but this is not the case with one's intellect. The more one uses one's mind, the sharper it gets. Of course one gets tired from overwork; but once rested, one's mind is improved, not dulled, as a result of the exertion.

354. This is an absurd comparison with the only thing in common in the two cases being the word "tear." Aztec kings took the lives of their victims (when they tore those victims' hearts out), but the Houston surgeons are involved in open-heart surgery in an effort to save lives.

357. One cannot compare a human being to a stone or a fire. They perform as they do as the result of the force of natural law on them. Human beings, having intelligence, have a measure of choice and therefore can be held at least partly responsible for their actions.

358. This is an attempt to compare "grafting" with "cannibalism." In both cases there is "assimilation" of someone else's flesh or organ into one's own body. One of the main differences is that in the case of grafting the organ in question continues to function as it had and is not transformed into energy and waste matter. (In addition, of course, in the case of grafting the donor has given his or her consent, which is not so in the case of cannibalism).

360. There is obviously a striking difference between a man's and a woman's chest, and it is not solely a matter of less hair and more fat. In our society and culture women's breasts are sexually arousing, but a man's hairy chest is not. Such toplessness might prove more troublesome (for all concerned) than liberating.

366. The existence of divots on a golf course is certainly a sign of the existence of certain kinds of players (and naming them is unnecessary); but the existence of certain types of, say, criticism or influence in Washington is by itself not sufficient proof of the existence of Communists. In the latter Welch is under an obligation to provide evidence. In short, there is no comparison between making divots on golf courses and making waves in Washington.

369. Waiters and hotel servants receive a lower wage because employers anticipate their employees will be tipped in relation to how friendly or helpful they are. Clerks, doctors, or bus drivers receive a wage or payment for a specified job, in which extending an extra friendly smile, say, or some other courtesy is not a necessary or relevant part of the work and tipping would be therefore inappropriate. Another way of putting this: in the case of waiters or hotel servants tipping stimulates them to be extra friendly and helpful; lawyers and doctors are not expected to try to earn extra compensation by extending such further courtesies.

False Cause

377. The chain letter purports to be the cause of good luck (if you follow its directions, or bad luck if you do not). A chain letter by itself cannot produce any such effects (although *believing* that it can indeed might). I have received some of these letters, which seem to have gone through quite a few hands, where the English and spelling are charming and quaint.

384. Even if it were the case that with the increase in welfare legislation there was a simultaneous increase in immorality, alcoholism, and suicide, this does not mean there is a causal relation between them. More than mere simultaneity would be needed to establish such a connection.

389.   The fact that her complexion cleared up soon after using that cream is not sufficient to claim that the cream did it. It might have been caused by other factors, such as a change in diet or frequent washing.

390.   Other factors (such as improved economic conditions or the season of year) may have been responsible for the increase in sales—although the advertising campaign may have played a part, too. There is not sufficient reason here to ascribe the sole part to it.

396.   Although it is no doubt a comfort to believe one's baldness is a result of gorgeous women running their fingers through one's hair, the cause is more likely the unfortunate genes one has inherited.

399.   Trying to attribute one's culinary habits and desires to traumas in early childhood, although understandable and tempting in this age of psychoanalysis, seems somewhat forced. More than likely there is only a verbal connection, not a causal one, between fearing "strange" foods and fearing "strangers."

400.   Cause and effect are reversed here: the sun setting causes the shadows, which then seem to blot out the sunlight—and not the other way around (night with its shadows causing the setting of the sun).

401.   The writer believes strong law enforcement actually causes crime because police patrols are most frequent in areas where there is a high crime rate. But here, too, the cause-effect relationship has been reversed. The high crime rate is the cause or reason for the frequent police patrols, not the other way around. If the police stay there long enough, the crime rate will go down.

405.   An increase in armaments is not the cause of war. Rather, war and an increase in armaments are usually the result of some other (third) cause. What one needs to ask here is why the sudden rush to arm. The answer to that question may bring us much closer to the actual cause of the war (if indeed it makes sense to search for some single cause where something as complicated as a war is involved).

406.   The reason for increased crime *and* religion in cities is common to both: larger populations than are to be found in rural communities. In other words, it is not that in larger cities religion causes crime but rather that social factors found in large cities are the reason for the existence of both.

411.   Cathy assumes the clerk is crabby because there is no bathroom the clerk can use. The clerk's grouchiness is actually caused by Cathy's refusal to believe her.

Irrelevant Thesis

414.   The fact that India is a great country is irrelevant to whether or not the administration is or was anti-Indian.

417.   The fact that killer-diller comic books are just the modern version of the dime novel is irrelevant to whether or not they hurt youngsters. The dime novels may also have had an adverse effect on youngsters.

421.   The host's conclusion is irrelevant in that it grossly exaggerates the guest's position; the fact that someone would like to see welfare benefits increased does not prove he or she wants the country to go Communist.

423.   The real question here is whether the practice of vegetarianism is injurious to those who want to be vegetarians. Certainly, it will prove "harmful" to the economy (it would eliminate the meat-packing business, for one); but the real question is whether it is "harmful" to the individuals who wish to limit themselves to a diet of vegetables.

424.   The speaker is asked to give to one, not to all. Therefore the fact that she cannot respond to all the charitable appeals is irrelevant to the question of why she cannot respond to this particular one.

426.   Communism may or may not have this consequence (and no doubt certain provisions could be made to avoid it); but even if it did, the objection would be irrelevant. The consequences of a certain (however bad or harmful) contention are irrelevant to its truth.

430.   Johnson's conclusion that by the time you try to resolve such a question some other boy has learned both is irrelevant to the original question (namely, which subject in fact should be taught first) and not an answer to it.

433. The fact that Meany does not like young people or that they smoke pot is no proof and is irrelevant to the question of whether they have anything to say. Listening to them does not mean they are going to control the destiny of the country. As in all the other cases of irrelevant thesis examined here, Meany does not face the question posed but deals with matters irrelevant to it.

TEST ON THE FALLACIES
OF PRESUMPTION

1. Advertisement for Bohemia beer:

If you had to name the three finest beers in the world, what would the other two be?

2. Saying:

Better safe than sorry.

3. Letter to the editor:

The way people dress is not an essential indication of the way they are or what they are. Despite the old adage, clothes do not always make the man. Some very "classily" dressed people are far from classy, and Einstein was not infrequently found strolling about with only one sock on. Besides, it is not true that we were different twenty years ago.

4. From a seventeenth-century religious circular:

Theology teaches that the sun has been created in order to illuminate the earth. But one moves the torch in order to illuminate the house, and not the house in order to be illuminated by the torch. Hence it is the sun which revolves around the earth, and not the earth which revolves around the sun.

5. From *Reader's Digest:*

A remarkable new idea that is picking up momentum in Washington could provide a solution to one of the nation's oldest governmental problems: how to force the federal bureaucracy to stop lying about its failings and start giving us an honest account of its doings. The idea is amazingly simple—pass a "whistle blower's law" that would give the bureaucrats themselves an opportunity to confess what they're doing wrong.

Excerpt from "What Happens When Bureaucrats Blow Whistles?" by James Nathan Miller, Reader's Digest July 1978.

6.

Without chemicals, life itself would be impossible.

Some people think anything "chemical" is bad and anything "natural" is good. Yet nature is chemical.

Plant life generates the oxygen we need through a chemical process called photosynthesis. When you breathe, your body absorbs that oxygen through a chemical reaction with your blood.

Life is chemical. And with chemicals, companies like Monsanto are working to help improve the quality of life.

Chemicals help you live longer. Rickets was a common childhood disease until a chemical called Vitamin D was added to milk and other foods.

Chemicals help you eat better. Chemical weed-killers have dramatically increased the supply and availability of our food. But no chemical is totally safe, all the time, everywhere. In nature or the laboratory. The real challenge is to use chemicals properly. To help make life a lot more livable.

Courtesy Monsanto Company.

7. Article by Anthony Cook:

It was his first visit to the Middle East and Dr. John Hubbard, the president of the University of Southern California, was sitting with a group of USC alumni—Arab alumni—at a banquet they were holding in his honor. After a sumptuous meal of lamb and pilaf, the group filed into an adjoining room for a little entertainment; filmed highlights of recent USC football seasons. The Arabs cheered as they watched the classic 1974 USC–Notre Dame game. The film showed USC down, 24 to 6, at the half, and the alumni groaned. Then their team came back, scoring 49 points to win it, 55 to 24. There were bursts of applause, and when the lights went on, the host rose and, filled with the spirit of victory, delivered a solemn pronouncement to the roomful of beards and robes, "Gentlemen," the host said, "Allah is a Trojan."

8. Speaker:

My opponent in this debate has argued that our prisons turn out more hardened criminals than enter them. He recommends emphasizing rehabilitative possibilities. He has mentioned job training, psychological therapy, and other education programs. Well, I certainly don't like all the prison riots and so forth that we've recently had, but I don't think we should change prisons into schools and hospitals. After all, criminals are in jail because they broke the law. They should pay the penalties.

9. **President Truman, regarding the cessation of atmospheric testing of nuclear weapons:**

We should never have stopped it. Where would we be today if Thomas Edison had been forced to stop his experiments with the electric bulb?

10. **Letter to the editor:**

It is not enough that the United States Congress has sloughed large portions of its responsibility to armies of anonymous bureaucrats aloof from the desires of the American public. Now, some of our elected representatives, apparently seeking to shirk the last vestiges of accountability to the people who elect and pay them, have voted to further stifle the voice of the people.

11. **News report:**

Drivers who drive alertly are good drivers because they are ready for anything. They think ahead. Safe drivers take precautions against the unexpected.

12. **DEAR ABBY:**

Recently I drove through a small "art-colony" village in Pennsylvania, which is normally frequented by tourists. I got the shock of my life when I saw about 75 young people all dressed exactly alike—in blue denim! I wondered if there had been a prison break, or an invasion of the "Union Army." What is it with our young people? They have about as much individuality as connected sausage links. They all look alike. Same dress, same jeans, same long straight hair—it's hard to tell one from the other. Go anywhere young people gather, high schools, colleges, social affairs, it's the same story. Why are they afraid to be different? It wasn't like that 20 years ago. Kids looked and acted like individuals and enjoyed it.

WONDERING WANDERER

Courtesy Chicago Tribune–New York News Syndicate, Inc.

13. **Sitcom:**

Son: Look, Mom. You divorced Dad because of his drinking, and now you're doing the same thing! What's wrong? Why do you do it?

Mom: Now dear. Your father was an *alcoholic!* I just take an occasional glass of wine. It relaxes me so I can face the terrible life I have to live. It's not fair of you to accuse me of such an awful thing!

14. **News report from the Associated Press:**

THE ACE OF SPADES
—A SIGN OF DOOM?

CONISTON, England, Jan. 4 (AP)—Donald Malcolm Campbell, who set speed records on land and water, was playing cards with friends last night.

He turned up the ace of spades, followed by the queen of spades and appeared upset.

"Mary, Queen of Scots, turned up the same combination of cards and knew from it that she was going to be beheaded," he told his friends.

"I know that one of my family is going to get the chop (killed). I pray to God it is not me."

This morning on Lake Coniston, his jet hydroplane Bluebird somersaulted and sank while traveling about 300 miles an hour. Campbell was killed.

15. **Cliché:**

A miss is as good as a mile.

16. **Popular maxim:**

Honesty is praiseworthy because it deserves the approval of all.

17. **Overheard:**

Why is it that girls are more interested in religion than are boys?

18. **Letter to the editor (regarding the cover picture on *Time* with the caption: "The Girl Who Almost Killed Ford" [Lynette "Squeaky" Fromme]):**

If a 27-year-old male had held the gun, would your headline have been "The Boy Who Almost Killed Ford?"

Grace Welch

Time, October 6, 1975, p. 3. Reprinted by permission from *Time, The Weekly Newsmagazine;* Copyright Time Inc. 1975.

19. **From Longley and Braun:**

"The problem of the winner losing and the loser winning in a democratic political system is a profound one. Professor Paul Freund of Harvard Law School aptly summed up this problem when he testified: 'The one objective that any democratic electoral system must achieve is to avoid the election of a candidate who secures fewer popular votes than an opposing candidate. The electoral college system offers no assurance of this, and in fact three times in our history the election went to a candidate other than the winner in the popular count. It has been said that this record is a good one, showing that in 93 percent of our elections the popular winner was the actual winner. This is like boasting that 93 percent of the planes leaving Washington airport arrive at their destination.'"

Lawrence Longley and Alan Braun, *The Politics of Electoral College Reform (New Haven, Conn.: Yale University Press, 1972),* pp. 20–21.

20. **Question:**

Isn't it true that students who get all A's study hard? So if you want me to study hard, give me all A's.

21. **Sign on Indian reservation:**

America: Love It or Give It Back.

22. **Posted sign:**

Notice to Our Customers: Because of continually rising costs for service, labor, repair parts, and other expenses, it has become necessary to increase the cost of washing to 25¢. You can be sure that this price increase will NOT affect the high quality of our service and machines. Thank you for your understanding and patronage.

23. Speaker:

Why is private enterprise so much more efficient than any government control of industry?

24. Letter:

Perhaps none of the accusations against Mr. Brown have been proved. But where there's smoke, there's fire. And if his record were really clear, all these charges wouldn't have been made against him.

25. From *American Rifleman:*

The value you place on a course that appeals to the nonshooting public depends entirely upon how much you value your freedom under the Second Amendment. Bear in mind that the antigun syndrome is an illness and that education is the prescription for a cure.

Bernie Fuller, "The Prescription," *American Rifleman,* May 1978.

26. From David Hume:

When we run over libraries, persuaded of these principles, what havoc must we make? If we take in our hand any volume; of divinity or school metaphysics, for instance; let us ask, *Does it contain any abstract reasoning concerning quantity or number?* No. *Does it contain any experimental reasoning concerning matter of fact and experience?* No. Commit it then to the flames: for it can contain nothing but sophistry and illusion.

David Hume, *An Enquiry Concerning Human Understanding* (1748) (Indianapolis: Hackett Publishing Company, 1977), p. 114.

27. Teacher:

According to Freud we are often frustrated because our sex drives are blocked; and they become blocked apparently because we are thwarted in our desires.

28. Overheard:

The wives of successful men wear expensive clothing, so the best way for a woman to help her husband become a success is to buy expensive clothing.

29. Letter to the editor:

Last week, we had lunch in a restaurant. Our waiter said, "Well, girls, have you decided what you would like to eat?" We then explained to him that we were no longer girls. We were women. He laughed and pointed to two older women at another table and said that he had even called them girls. Later, still infuriated, we called him "boy." He went off in a rage. We tried to explain to him that he was as much a boy as we were girls. He didn't understand.

30. From *Time:*

A favorite word of the Vatican II years was "collegiality," referring to power sharing between the Pope and the 2,280 bishops across the world. Paul took the dramatic step of forming an international Synod of Bishops, then made it a forum without power. It met five times during his reign and never had any discernible effect on his thought or action. . . . Father Herbert Ryan, a Los Angeles theologian, argues that Paul had good reason to distrust power sharing. The bishops counseled him privately that the explosive *Humanae Vitae* birth control encyclical would win gradual acceptance, and they were proved spectacularly wrong.

ANSWERS TO TEST
ON THE FALLACIES
OF PRESUMPTION

1. Complex question
2. Bifurcation
3. Hasty generalization
4. Begging the question
5. Question-begging epithet
6. Sweeping generalization
7. Hasty generalization
8. Irrelevant thesis
9. False analogy
10. Question-begging epithet
11. Begging the question
12. Hasty generalization
13. Special pleading
14. False cause
15. Bifurcation
16. Begging the question
17. Complex question
18. Special pleading
19. False analogy
20. False cause
21. Bifurcation
22. Irrelevant thesis
23. Complex question
24. Sweeping generalization
25. False analogy
26. Bifurcation
27. Begging the question
28. False cause
29. Special pleading
30. Hasty generalization

PART THREE

Fallacies

Of Relevance

A Cure for TV Violence

Personal Attack

The first of the fallacies generally discussed under relevance is *personal attack*. This fallacy assumes a number of different forms. The first is *genetic fallacy*, a type of argument in which an attempt is made to prove a conclusion false by condemning its source or genesis. Since it is irrelevant how an idea originated, the argument is fallacious. Exercises 434 to 438 are examples of this form of personal attack.

Genetic Fallacy

*434. Overheard:

Religion began with magic and animism. Religion is therefore nothing but nonsense.

435. From a logic textbook:

Humans are made of nothing but atoms; atoms have no free will, therefore humans have none either.

436. Letter:

It must be false that in a capitalistic economy small business tends to disappear because this is said in the *Communist Manifesto*, and the only person I have ever heard affirm this was a Communist.

437. Overheard:

This piece of legislation is calculated to exploit the poor, for it was written and sponsored by the richest person in the state.

*438. From George Santayana:

Nietzsche was personally more philosophical than his philosophy. His talk about power, harshness, and superb immorality was the hobby of a harmless young scholar and constitutional invalid.

George Santayana, "Egotism in German Philosophy" (1915), in *The German Mind: A Philosophical Diagnosis* (New York: Harper & Row, Pub., 1968), p. 127.

Abusive ad Hominem

Abusive ad hominem differs from genetic fallacy in that besides attempting to dispose of a person's argument by condemning its source, the person showers the source with abuse. It is an effective tactic since by lowering a person in our sight we also succeed in lowering our estimation of whatever he or she has to say. There are many ways we can accomplish this: we may try to raise suspicions about a person (as in number 439), try to make a person look ridiculous (440), show our contempt (441 to 445), or charge the person with being inconsistent (446 to 448). There are no doubt other ways as well, but these are typical.

439. Overheard:

This theory was introduced by a person of known Communist sympathies. There cannot be much to it.

440. Speaker:

We shall reject Mr. Jones's suggestion for increasing the efficiency of our colleges. As a manufacturer, he cannot be expected to realize that our aim is to educate the youth, not to make a profit.

441. Editorial:

Present economic policies are rapidly placing this country in a bad condition. This is mainly due to some of the ex-White House advisers connected with the former administration plus some of the eggheads still in power. These people are evidently very egotistical and smug, entirely out of contact with the American people and Congress.

*442. From Bernard Bykhovsky:

Mr. North, a fascist flunkey, wishes by means of his semantics to make meaningless not only the sacred deeds of heroism of the fighters against fascism, but the whole past and future struggle for liberty. But he only exposes his reactionary guts, his hatred for liberty and for social progress....

As a result of Mr. Stuart Chase's many years of activity in the American press, his ability to understand "truth" and "national honor" must have become atrophied; as for his ability to understand a "classless society," it is inherent with him....

Stuart Chase, the petty bourgeois American economist, who writes prescriptions for the diseases of capitalism, having read the writings of semanticists has lost the last remnants of common sense and has come forward with a fanatical sermon of the new faith, a belief in the magical power of words.

Bernard Emmanuilovich Bykhovsky, *The Decay of Contemporary Bourgeois Philosophy* (Moscow: Mysl, 1947), p. 173.

443. Interchange:

Jones: What you now say passes my poor powers of comprehension; it may be all very true, but I can't understand it, and I refrain from any expression of opinion on it.

Smith: I beg your pardon; but, with your penetrating intellect, it must be very easy for you to understand anything; and it can only be my poor statement of the matter that is at fault.

444. Speaker:

We need not consider this piece of social legislation. It was, as you know, introduced by Senator Farrell, who is just not a very socially enlightened person.

445. From a magazine:

America has always been able to find jobs for the unskilled. Look at all the people we've sent to Congress.

Courtesy *The Comedy Center*, Wilmington, Delaware.

*446. Speaker:

He is an American and thus should believe that Americans make the finest soldiers in the world.

447. Speaker:

She is a doctor of medicine and therefore should condemn the profession of chiropractic.

*448. Speaker:

> As a manufacturer you should have supported this bill urging higher tariffs!

Circumstantial ad Hominem

In the circumstantial ad hominem fallacy an attempt is made to undercut opponents' positions by suggesting that in advancing their views they are merely serving their own interests. Exercises 449 to 454 are some typical examples of this form of the fallacy.

449. Speaker:

> This rent control bill is unworkable and unjust, for Mr. Jones and those who have joined him in sponsoring it are all tenants and renters; there isn't a single landlord in the group.

450. Overheard:

> The president of the bank maintains that personal income taxes for the wealthy should be reduced. It is just what you would expect from a person who has a big income and is greedy for more.

451. Interchange:

> Smith: Of course you would be in favor of reduced real estate taxes because you would benefit personally by such a reduction.
>
> Jones: Of course you are against such a reduction because you own no real estate.

*452. Letter to the editor:

> One cannot believe the arguments of conscientious objectors, since they are obviously trying to escape the draft.

453. Letter to the editor:

> The senator is really sponsoring the amendment primarily as a way to gain national recognition in his bid for the presidency.

454. From *Time* magazine:

Murray Weidenbaum: Somehow Congress got sold on the notion of "pay comparability" between the public and private sectors, ignoring the high federal fringes. And who makes the computations of the "comparability"? Surprise, surprise! It's the civil servants themselves, which is like having the foxes guard the henhouse.

May 28, 1978, p. 76. Reprinted by permission from *Time, The Weekly Newsmagazine;* Copyright Time Inc. 1978.

Tu Quoque

The traditional label that the *tu quoque* form of the fallacy of personal attack carries has no simple translation. A colloquial equivalent in English would be, "Hey, look who's talking!" (or "You too!"), as in the following interchange:

Smith: Smoking is unhealthy. You should quit.
Jones: Hey! Look who's talking! You smoke!

This is a fallacy, of course, because the fact that Smith smokes does not invalidate his contention that smoking is unhealthy. The *tu quoque* generally takes two forms: attacking the other person for acting in a manner that contradicts the very position he or she advocates (exercises 455 to 461); or attempting to defend oneself for engaging in the particular questionable activity by arguing that the other person engages in it too (exercises 462 to 473). In either case, since two wrongs do not make a right, the arguments are fallacious.

455. Overheard:

If the Peace Corps is so great, why aren't you in it?

*456. Letter to the editor:

I find it incredible that Turkey, which has been one of the world's producers of opium for heroin, would sentence three Americans to life imprisonment for supposedly smuggling marijuana into Turkey. This is surely a travesty of justice and an act of hostility toward the American people.

457. **News report:**

A top Soviet authority on U.S. affairs showed irritation at Americans over the humanitarian issue. "What moral right do they have to act as preachers of freedom and democracy, especially in the light of events which occurred in recent years in America itself?" asked Georgy Arbatov, listing Watergate and Wounded Knee.

458. **Speaker:**

Georgia Republicans say they don't know of a single black Republican lawyer who could be appointed to a judgeship. They add smugly that if the Democrats couldn't find a Negro judge in all those years how can anybody expect the Republicans to find one now?

*459. **Novelist Ira Levin:**

Your reviewer charges me with bad taste in using Dr. Josef Mengele, late of Auschwitz, as the villain of my novel *The Boys from Brazil* (February 23). I must concede that what I have done is almost on a par with putting a would-be assassin on the cover of a national magazine or publishing a list of a dead president's rumored mistresses.

*460. **From a book on Mohammed:**

Undoubtedly a prophet who declares that women and children belong to the enticements of worldly life, and who nevertheless accumulates a harem of nine wives, in addition to various slave women, is a strange phenomenon when regarded from the standpoint of morality. The situation is not improved by the fact that up to Khadijah's (Mohammed's first wife) death, that is, until Mohammed was fifty years old, he was content with one wife. At the height of his career, when he was already an ageing man, he gave free rein to his sensual impulses.

Tor Andrae, *Mohammed, The Man and His Faith,* trans. Theophil Menzel (New York: Harper Torchbooks, 1960), p. 188. By permission of Harper & Row, Publishers, Inc.

461. From Hume:

But I observe, says Cleanthes, with regard to you, Philo, and all speculative sceptics that your doctrine and practice are as much at variance in the most abstruse points of theory as in the conduct of common life.

David Hume, *Dialogues Concerning Natural Religion* (1779). (New York: Hafner Press, 1959), p. 10.

462. Interchange:

"Son, why are your grades so low?"
"Father, did you ever make a low grade in school?"

463. Report:

The doctor spoke tactfully to a patient about his bill, remarking, "I don't like to bring this up, but that check of yours came back." "I don't like to mention this, either, doctor," said the patient, "but so did my gout."

464. News item:

Ilie Nastase, asked if he ever has second thoughts about his behavior on the court: "No. Why should I? The crowd doesn't have second thoughts for me. And when I lose, they're happy afterward."

Los Angeles Times, May 21, 1978. Copyright 1978, *Los Angeles Times.* Reprinted by permission.

*465. Speaker:

It is strange that you should think it inhuman of me to take so much pleasure in hunting; you don't seem to mind feeding on the flesh of harmless animals.

466. Speaker:

My client did not act improperly in using an official car for commuting between his home and his office. Even the mayor does this from time to time.

467. **Former mayor of Chicago, Richard Daley, on government in-fallibility:**

Look at our Lord's disciples. One denied him, one doubted him, one betrayed him. If our Lord couldn't have perfection, how are you going to have it in city government?

*468. **From Benjamin Franklin:**

I believe I have omitted mentioning that in my first voyage from Boston to Philadelphia, being becalmed off Block Island, our crew employed themselves catching cod and hauled up a great number. 'Till then I had stuck to my resolution to eat nothing that had had life; and on this occasion I considered, according to my Master Tryon, the taking every fish as a kind of provoked murder, since none of them had or ever could do us any injury that might justify this massacre. All this seemed very reasonable. But I had formerly been a great lover of fish, and when this came hot out of the frying pan, it smelled admirably well. I balanced some time between principle and inclination till I recollected that when the fish were opened, I saw smaller fish taken out of their stomachs. "Then," thought I, "if you eat one another, I don't see why we mayn't eat you." So I dined upon cod very heartily and have since continued to eat as other people, returning only now and then occasionally to a vegetable diet. So convenient a thing it is to be a *reasonable creature*, since it enables one to find or make a reason for everything one has a mind to do.

The Autobiography of Benjamin Franklin, ed. Russel Nye (Boston: Houghton Mifflin, 1958), pp. 31–32.

*469. **Speaker:**

Peaceful demonstrations just don't work. Whatever the Left may do is not as violent as that of the Establishment.

470. **Overheard:**

When you've got a better theory, you can criticize mine.

471. **Overheard:**

Sure my theory is inconsistent, but you have no right to criticize it because you don't even have a theory.

472.   Overheard:

It is not dishonest. Why, if you were to get too much change, you would probably keep it too.

*473.   Speaker:

The socialists would do just what the capitalists do if they had the chance.

Poisoning the Well

The tactic here is somewhat more complicated. It involves attempting to place one's opponent in a position from which he or she is unable to reply. Its purpose is therefore to avoid opposition by precluding discussion (as in number 474). Sometimes the very attempt to reply succeeds only in placing one in an even more impossible position—as in the case where one might say of someone that he talks too much. Should such a person try to defend himself against the charge, he would only succeed in confirming the contention, allowing us to say, "See, I told you!" (Number 475 is a more serious variant of this device.) As the remaining examples of this form of the fallacy (exercises 476 to 481) show, such unfair tactics are designed to discredit in advance the only source from which evidence for or against a particular position can arise.

*474.   Cliché:

Don't listen to him; he's a liar.

475.   Speaker:

We must assume she is a Communist because she says she is not and we know Communists are taught to deny Party membership.

*476.   Cliché:

If you disagree with me, you don't understand what I'm saying.

477.   Speaker:

Your attempt to show that belief in democracy is not simply an emotion of fear is itself due to an emotion of fear, and you thus verify rather than refute my thesis.

*478. Speaker:

The person who rejects my conclusions is either an insane bigot who claims my pity or a foul-mouthed slanderer who has my contempt.

*479. News report:

Sen. William Scott (R.-Va) denied yesterday that he was "the dumbest" member of Congress, as *New Times* magazine said he was, but, he added, he was in a quandary whether it would be wise to take the matter to court.... He hasn't decided yet whether to file a libel against *New Times,* Scott said, because in libel suits the element of malice, a crucial point, is very hard to prove. "I don't want to bring a suit that I don't win," he said.

"Sen. Scott Won't Test His Intelligence," *Miami News,* May 24, 1974. Courtesy *Miami News,* Miami, Florida. Taken from Howard Pospesel and David Marans, *Arguments: Deductive Logic Exercises,* 2nd ed. (Englewood Cliffs, N.J.: Prentice-Hall, 1978), p. 46.

480. News report:

Zsa Zsa Gabor was told in front of millions of television viewers last night she is vain, untalented and a complete non-event.... The incident took place on a late-night celebrity program, "The Eamonn Andrews Show." Zsa Zsa's outspoken fellow guest was comedian Peter Cook.... "You cannot be very talented yourself otherwise you would recognize talent in others and you would not have said I was untalented," bit back Miss Gabor (in response to Cook's criticism).

"Zsa Zsa Told She's Untalented Non-Event," Reuters News Service, *Miami News,* January 10, 1969, p. 5-A. Courtesy Miami News, Miami, Florida. Taken from Howard Pospesel, *Arguments: Deductive Logic Exercises* (Englewood Cliffs, N.J.: Prentice-Hall, 1971), pp. 195–96.

481. From *Hidden Meanings:*

"I don't think I really matter to you."
"Now why are you saying that? I'm doing the best I can."
"Well, I just feel taken for granted."
"I think you are insatiable. There is never enough."

"See, this is proof of what I just said. I don't really matter to you. If I did you wouldn't talk this way to me."

Gerald Walker, *Hidden Meanings* (Millbrae, California: Celestial Arts, 1977), pp. 134–35.

482. **From R.D. Laing, H. Phillipson and A.R. Lee, *Interpersonal Perception: A Theory and a Method of Research* (New York: Springer Publishing Company, Inc., 1966), pp. 31–33:**

"Jack persistently refuses to infer from Jill's behavior towards him, however loving, that she "really" loves him, but believes, despite evidence from Jill's manifest behavior . . . that she loves Tom, Dick or Harry. A curious feature of Jack's tendency to attribute to Jill a lack of love for him and a love for Tom, Dick or Harry . . . often seems to be that he tends to make this attribution in inverse proportion to Jill's testimony and actions to the contrary.

Jack may reason: "Look at all the things that Jill is doing to try to prove to me that she loves me. If she really loved me she would not have to be so obvious about it and try so hard. The fact that she is trying so hard proves she is pretending. She must be trying to cover up her feelings—she must be trying to cover up her true feelings. She probably loves Tom."

At this point Jill is in a double-bind. . . . If she tries to act even more loving, she further activates Jack's assumption that she is pretending. If, on the other hand, she pretends to act less loving and more aloof then she certainly will activate his view that she does not love him. He then can say: "See, I told you so, she really doesn't love me. Look at how aloof she has become. . . ."

Now, Jack may decide to resolve his mistrust by various moves that one generally regards as part of the paranoid strategy. He may pretend to Jill that he does think she loves him, so that, in his view of her, she will think she has fooled him. He will then mount evidence (she has exchanged glances with a man, she smiled at a man, her walk gives her away because it is the way a prostitute walks, etc.) that seems to him to substantiate his secretly held view that she does not love him. But as his suspicion mounts, he may discover that the evidence he has accumulated suddenly looks very thin. This does not prove, however, that his attribution is correct; it proves that he has not taken into account how clever she is. . . . Thus he reasons: "I have not been smart enough. She realizes that I am suspicious so she is not giving anything away. I had better bluff her by pretending to some suspicions that I do not feel, so that she will think I'm on the wrong track." So he pretends to her that he thinks she is having an affair with Tom, when he "knows" that she is having an affair with Dick.

Used by permission.

Mob Appeal

The lesson of the *mob appeal* fallacy is that one's allegiance is too important to be given lightly at the flashing of a sign or the ringing of a familiar bell, that we need to temper our high-mindedness with a dash of cold reason. We need to try to desensitize ourselves, it teaches us, to the highly inflamed rhetoric with which we are now constantly bombarded, not only from politicians trying to get our vote or hate-mongers out to contaminate us with their hate but even from well-known and respected business firms out to get our dollars and therefore busy promoting their products in less than strictly honest ways. Their devices are to stir up our emotions and short-circuit our thinking by using highly explosive and emotional words in the hope that we will substitute feeling for logical thinking. Exercises 483 to 494 are examples of these devices of mob appeal: 483 to 486 are examples of the use of the device in advertising and 487 to 494, of its use in politics and hate literature.

*483. **Advertisement by Licensed Beverage Industries, Inc.:**

One American custom that has never changed: a friendly social drink.

484. **Seagram advertisement on the occasion of Ray Milland winning an Oscar for his portrayal of Don Birnam in the movie version of Charles Jackson's novel *The Lost Weekend* (1944)—one of the first novels (and movies) to deal compassionately and realistically with the problem of alcoholism:**

The House of Seagram congratulates Ray Milland
on his magnificent performance
in
The Lost Weekend

485.

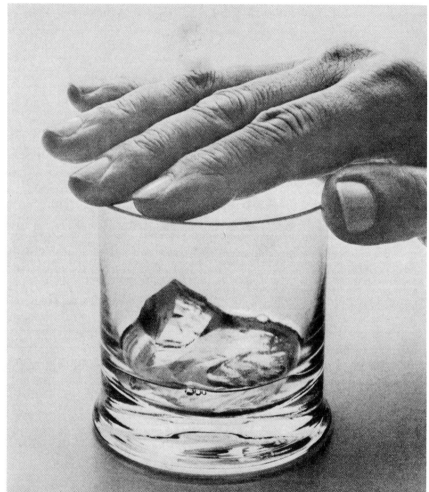

Your guest is trying to tell you something. Please listen.

The good host serves more than food and drink. He serves his guest.
By giving him his attention. By making him feel comfortable. By listening to what he wants...and doesn't want.
Next time your guest decides he's had enough, be a good enough host to take him at his word...or his sign. He'll think better of you for it.

Seagram
Distillers Company.

486. Advertisement for Pan Am Airlines:

AMERICA'S AIRLINE TO THE WORLD

Courtesy Pan American World Airways.

487. Letter to alumni:

This is a call to action—a call to you to join the great Trojan alumni team which is going to make this year another number one season in support of USC! On campus the 1973 school year started with terrific momentum—the fall influx of notable scholars who make USC both prominent and dominant.

We alumni who have benefited from USC should give something to replant the seed, to help keep our university growing in the practical, energetic, and superior way you and I know and love. Let's raise the sword in the hand of Tommy Trojan just a little higher this year in salute to alumni power. Your immediate alumni fund contribution will give the opportunity to our present and future Trojans to enjoy and benefit the same way you and I have in being a Trojan. I've made the move, how about you!

488. Speaker:

No one in this room wants to deny any child a decent education. But let's remember that this is our school and it belongs to our children.

*489. Circular:

But Americanism Educational League will NEVER waver! We know—and you know—that our economic and governmental systems have brought freedom from want, freedom from fear and freedom of opportunity to more people over a longer time than any system in history. We—and you—know that, in spite of temporary setbacks, America is and will remain the greatest country on God's earth!

You can have a vital part in our patriotic campaign on behalf of the Real America...the land of brave, proud, confident, hardworking people. Your tax-deductible contributions can multiply our efforts. You can help us guide our youth along good pathways and fight the one-world socialists who would sell us out. We need your personal interest, your support of our many projects, your financial help ... and your prayers!

490. Speaker:

My candidate should be elected. Or should we be governed by a political opportunist who has cleverly ridden into office after deliberately arousing a mass hysterical reaction?

*491. Letter to the editor:

In answer to Louis Archer who states we should send our troops overseas to capture Arab oil fields and that he, for one, is "ready to go." It seems the war-hungry are now developing an unholy thirst for oil. To him and others of his ilk, let me say, your thirsts will never be quenched—nor your gas tanks filled—with my sons' blood.

492. Circular from the National Socialist White People's Party:

No Troops for Israel!
Americans Will Not Die
For the Jews!

Are you tired of the Establishment's nauseating slavery to the Jews? Are you weary of seeing America support a gang of vicious murderers in the Middle East? Are you sick of political hacks spending your tax dollars on genocide for Arabs abroad and genocide for White people at home? Are you fed up with integration, pollution, treason in the streets, forced busing, nigger crime, and Watergate-style corruption in high places? Are you ready for a new kind of politics, for a political party which stands up for the interests of the *White* majority in America?

493. Another circular from the National Socialist White People's Party:

Had enough, "Pig"?

Had enough of being called "Pig" by scum not fit to wipe your feet on?

Are you sick and tired of hearing about "police brutality" from the swarms of degenerates in the streets—and from City Hall and Congress too?

Are you getting a little tired of seeing Black criminals back on the streets a few hours after you've arrested them for serious crimes?

Are you fed up with risking your life for a public which is too gutless to back you up politically and too ungrateful to pay you a decent salary?

You are expected to let the Reds and Negroes curse you, spit on you, throw bricks and bags of excrement at you, and still keep your cool. If you so much as crack a woolly head with your nightstick, you're liable to find yourself facing a civilian review board or a disciplinary committee. And if you use your firearm to give some looter, rioter, or other criminal what he really deserves, then you've really had it!

Don't you agree that the situation in American cities has gotten bad enough to call for drastic action to curb Black crime and Red demonstrations?

Well, brace yourself, Whitey, because it's going to get a lot worse.

In many major American cities the politicians are talking about disarming the police altogether in an effort to appease the cities' Negroes. And soon many White policemen will find themselves paired off on their beats with a Black. When one of these Blacks has to make a choice between his White partner and some "soul brother," it may be a bullet in the back for another White policeman.

And don't count on the Democrats, the Republicans or any of the "System" politicians for support. They may talk about "law and order" when election time rolls around, but we've had plenty of chance to see how little their promises are really worth.

You, better than most other citizens, KNOW what this country is headed for in the years to come. You KNOW what the street mobs have in mind for every "pig" who stands between them and taking control of this country. You KNOW, from everyday experience, what a pack of lies all this propaganda about "equality" and "human dignity" really is. You KNOW what the Black is really like. And you KNOW what the only REAL CURE is for those who are out to destroy our race and our nation.

What you've seen up till now is just the beginning of the hell that's going to break loose in the next few years. Let's begin doing something about it—NOW. Join us today!

494.   **From *Time* magazine:**

"A tide of color threatens to engulf Britain." So warns the National Front, a neofascist party whose main goal is to expel the estimated 2 million "coloreds"—Jamaicans, Indians, Pakistanis and other nonwhite former colonials—who have migrated to Britain since 1945. The ten-year-old front mixes crude, inflammatory racism with a dose of ultranationalism (calling for increased defense spending and high protective tariffs, for example). Official membership is only about 20,000, but the front has attracted a following among working-class whites and is the country's fastest-growing political movement. Although it has yet to elect a Member to the House of Commons, the front gained nearly 10% of the popular vote in a recent parliamentary by-election, trailing Labor and the Tories but nudging the venerable Liberals from third place.

The roots of the racial violence are familiar: a suspicion of unfamiliar customs and a tendency to blame problems on newcomers. Many Bradfordians complain that destitute immigrants are given cash payments by the government when they disembark at Heathrow Airport, while white pensioners must scrimp along on inadequate retirement pay. Describing herself as opposing immigration, Shopkeeper Patricia Barrow is convinced that the Pakistanis inhumanely slaughter sheep and chickens according to Islamic ritual; she also fears that "colored boys will be hanging around white girls."

Disgusted by the inability of the established parties to reduce inflation (now 14%), unemployment (1.5 million) and high taxes, thousands of Britons unfortunately find the front a refreshing, sympathetic alternative. Boasts James Merrick, 45, the party's spokesman in Bradford: "Sometimes people call us up and talk for hours. There is no other place for them to turn. They're just bursting with frustration." The Very Rev. Brandon Jackson, an Anglican priest in Bradford who opposes the front, reluctantly concedes the front's appeal: "People are so fed up with politics that they're impressed by anyone who'll take a stand. The front says in public what people say in private. It trades on people's fears."

Merrick insists that the coloreds are responsible for the country's chronic unemployment and rising lawlessness. Although police deny that immigrants are the major cause of increased crime, Merrick maintains that "coloreds accost half the women in town" and push drugs—which, he speculates, are cached in mosques and temples....

"The Coloreds Must Go!" *Time,* December 12, 1977, p. 50. Reprinted by permission from *Time, The Weekly Newsmagazine;* Copyright Time Inc., 1977.

Appeal to Pity

Like mob appeal, *appeal to pity* is designed to win people over by playing on their emotions. It is distinct from mob appeal in that it tries to short-circuit our thinking by exploiting one single emotion, that of sympathy.

Like mob appeal, this tactic has not been overlooked by advertisers; and the first three examples in the exercises that follow (495 to 497) are taken from that source. Another rather rich source of examples of the exploitation of this fallacy is our law courts, and several of the other examples come from that source. A well-known but absurd example of the latter type is the case of the juvenile delinquent who had murdered his mother and father and pleaded for clemency on the ground that he was an orphan.

We need to remember, however, that an appeal to emotion is not always irrelevant. For example, although it would be fallacious for a defense attorney to offer evidence about her client's unfortunate lot as a reason for the court to find him innocent of a crime of which he stands accused, it would not be fallacious for that attorney to offer such evidence as a reason for treating the accused with leniency. And needless to say we should not thumb our nose at emotion as such. Emotions are an important part of our makeup: were we not capable of becoming angry at injustice, it would not be corrected; were we not capable of feeling outraged at the oppression of the innocent, they would continue to be oppressed. Still we need to recognize that feelings often stand in the way and how they do so.

*495.

SWING BY HOME ON YOUR NEXT BUSINESS TRIP.

496.

"Life passes you by so fast, the least you can do is enjoy the trip."

FORD

When America needs a better idea, Ford puts it on wheels.

A BEAUTIFUL EXPERIENCE. MUSTANG II

Courtesy Ford Division.

497.   Advertisement for corn oil margarine:

Should an 8-year-old worry about cholesterol?

498.   Letter to the editor:

Permitting increased immigration will actually strengthen this nation's economy. The unhappy peoples of other lands, the homeless of the world, children torn from their mother's arms, should be given a chance to find a new life among us.

*499.   Defense attorney:

My client is the sole support of his aged parents. If he is sent to prison, it will break their hearts, and they will be left homeless and penniless. You surely cannot find it in your hearts to reach any other verdict than "not guilty."

500.   Overheard:

An attorney says a sixty-two-year-old man accused of bilking several members of the Seventh Day Adventist Church out of thousands of dollars could suffer a fatal attack if compelled to stand trial.

501.   Source unknown:

Have they (the family of the accused) any rights? Is there any reason, Your Honor, why their proud names and all the future generations that bear them shall have this bar sinister written across them? How many boys and girls, how many unborn children, will feel it? It is bad enough however it is. But it's not yet death on the scaffold. It's not that. And I ask Your Honor, in addition to all that I have said, to save two honorable families from a disgrace that never ends, and which could be of no avail to help any human that lives.

502.   Excerpt from *Clarence for the Defense* by Irving Stone:

I appeal to you not for Thomas Kidd, but I appeal to you for the long line—the long, long line of despoiled and downtrodden people of the earth. I appeal to you for those men who rise in the morning before daylight comes and who go home at night when the light has faded from the sky and give their life, their strength, their toil to make others rich and great. I appeal to you in the name of those little children, the living and unborn.

Clarence for the Defense (New York: Doubleday, 1941), p. 112. Copyright 1941 by Irving Stone. Reprinted by permission of Doubleday & Company, Inc.

*503.   Letter:

Dear Friend:

Streaks of pain etching his youthful face, the boy drags his withered legs along the sidewalk, two crutches biting into the pavement one short step ahead. The minutes seem like hours, but the thinly built boy finally makes it to his destination, hangs up his coat and prepares for his day's activities.

When a handicapped boy pains himself by walking grueling distances on hard wooden crutches, it proves that the handicapped will go out of their way to improve their lives. Now we ask you to go a little out of your way to help the handicapped.

Chairperson, Contributions Committee

Appeal to Authority

The *appeal to authority* fallacy arises from citing someone who is not an authority (an expert or a specialist) on the subject in "proof" of a position. "Misappeal to authority," as has often been pointed out, would no doubt be a more appropriate label for this fallacy.

As in the case of many of the other fallacies illustrated thus far, this fallacy can appear in a number of different forms. It can take the form of claiming something is so because so and so (an authority whose expertise lies elsewhere) said so (exercises 504 to 512). (This form of the fallacy seems the converse of the fallacy of personal attack: *it must be so because so and so said it was* versus *it cannot be so* because look who says it is.) It can take the form of claiming something is so because of the great numbers of people who have believed or claimed it to be so (exercises 513 to 524). This form of the appeal seems an especially favorite one with advertisers—especially in cigarette and liquor ads. It can also take the form of the appeal, not to the one or to the many, but to the select few (exercises 525 to 529). This form of the fallacy is sometimes called *snob appeal* (to contrast it, perhaps, with the fallacy of mob appeal). Finally, it can take the form of an appeal to the force of custom or tradition (exercises 530 to 534). This last form, too, has not been overlooked by advertisers, who recognize its enormous appeal.

*504. Logic text example:

No federal aid should be provided for public schools. My banker, my dentist, my doctor, my minister, and all my business associates are opposed to it.

*505. Overheard:

There must be something to psychical research; such famous physicists as Lodge and Jeans and Eddington took it seriously.

506. Overheard:

At noon I lunched with a friend, who told me I was only kidding myself. "Meat is more fattening than potatoes," he said. "My doctor lets me eat all the potatoes I want—without butter, of course." I have always respected this friend's judgment (he's made a million dollars in the stock market), so I had potatoes with my meat.

*507. Letter to the editor:

The hullabaloo over dishonesty among athletes at the all-state conference is unfounded. The mayor, who attended that conference, declared unequivocally that he saw no evidence of dishonesty.

508. From John Strachey:

In that melancholy book, *The Future of an Illusion*, Dr. Freud, himself one of the last great theorists of the European capitalist class, has stated with simple clarity the impossibility of religious belief for the educated man of today.

John Strachey, *The Coming Struggle for Power* (New York: Modern Library, 1935), p. 170.

509. From Henry Fielding:

The elegant Lord Shaftesbury somewhere objects to telling too much truth: by which it may be fairly inferred, that, in some cases, to lie is not only excusable, but commendable.

Henry Fielding, *Tom Jones* (1749) (New York: Random House, 1950), p. 645.

*510. From Galileo (the following is spoken by one of the participants, an interlocutor named Salviati):

But can you doubt that air has weight when you have the clear testimony of Aristotle affirming that all elements have weight including air, and excepting only fire?

Galileo Galilei, *Dialogues Concerning Two New Sciences* (1638), trans. Henry Crew and Alfonso De Salvio (New York: Macmillan, 1914), p. 77.

511. Advertisement:

My darling husband is a very demanding man. He insists on a good Sunday breakfast and Maxim. It tastes like perked coffee.

Courtesy of General Foods Corporation, owner of the registered trademark MAXIM.

512. **Advertisement:**

If you like people, be sure you brush with Colgate. Walt Frazier wouldn't think of brushing with anything else.

513. **Popular belief:**

In all times and places, in every culture and civilization, people have believed in the existence of some sort of deity. Therefore a supernatural being must exist.

*514. **Overheard:**

Everybody's wearing it.

*515. **Billboard:**

Eat American Lamb. Ten Million Coyotes Can't Be Wrong.

516. **Advertisement:**

What! You still don't own any mutual funds?

517. **Advertisement:**

99 new homeowners out of 100 heat with gas.

*518. **Advertisement:**

Over one billion people have flown Boeing 727s all over the world. No other jetliner has carried so many passengers. We asked 80,000 people in 90 countries why they preferred the 727. Among the reasons given were: room for carryons, pleasant interiors, not too big and easy on-easy off. More than 70 airlines now fly 727s. To nearly every major city in the world. Next time, take off on the No. 1 trijet.

519. **Advertisement:**

How 2,000 Children Proved
the Leading Fluoride Could *Not* Beat
Aim at Fighting Cavities.

520. Advertisement:

America Loves

BURGER
KING

Courtesy Burger King Corp.

521. Advertisement:

More people use Desenex to help stop Athlete's Foot than any other remedy.

Courtesy Pharmacraft Division, Pennwalt Corp.

522. Advertisement:

No one wants the evening to end when there's friendship, laughter and Cointreau, the world's largest selling liqueur spécialité.

Courtesy Cointreau, Ltd.

523. Advertisement:

Gordon's Gin. Largest seller in England, America, the world.

Courtesy Gordon's Dry Gin Company Limited.

524. Advertisement:

Gordon's didn't get to be the world's largest selling Gin without making an exceptional Martini.

Courtesy Gordon's Dry Gin Company Limited.

525. Television star:

Darling have you discovered Masterpiece. The most exciting men I know are smoking it.

526. Advertisement:

People who have everything always smoke Richman cigars.

527. Advertisement:

The Lancia concept.

**TO SET YOU APART
FROM THE CROWD.**

Since 1906, Lancia has pioneered one automotive innovation after another and made performance an integral part of design. Test drive this exceptional automobile. If you like traveling apart from the crowd, you'll like traveling in Lancia.
See your authorized Lancia dealer.

Lancia. The intelligent alternative.

Courtesy Lancia of America.—Division of Fiat Motors of North America, Inc.

*528. Advertisement for Pinch 12-year-old scotch:

Look at it this way: You figured you could save $5.50 a month by driving to work. Then you bought a $200 Citizens' Band Radio to make the trip fun. And you're still drinking ordinary Scotch?

Courtesy Renfield Importers.

529. Advertisement:

You can take a White Horse anywhere.

Courtesy White Horse Distillers Limited.

530. Advertisement:

Drink the Irish Mist. Ireland's legendary Liqueur.

Courtesy Heublein, Inc.

*531. From a logic text:

The institution of marriage is as old as human history and thus must be considered sacred.

532. From Justice Douglas:

England's King George III was the symbol against which our founders made a revolution now considered bright and glorious. We must realize that today's Establishment is the new George III. Whether it will continue to adhere to his tactics, we do not know. If it does, the redress, honored in tradition, is also revolution.

William O. Douglas, *Points of Rebellion* (New York: Random House, 1970), p. 95.

*533. Advertisement for Cutty Sark Scots Whisky:

AFTER 280 YEARS OF
DEALING WITH ROYALTY,
WE'VE LEARNED A LITTLE
SOMETHING ABOUT TASTE.

Courtesy The Buckingham Corporation.

534. Advertisement:

Ballantine's Scotch was there.

Harvard-Yale! 1934

A football Saturday in New Haven, 1934. This was it. The last game of the season. Harvard–Yale!

Chanting "Boola, Boola!," they spill out of fraternity houses and run for the stadium.

It was at such a moment that a coach had told his team, "Gentlemen, you are about to play football for Yale against Harvard. Never in your lives will you do anything so important."

Now, there is no stopping them. These are Saturday's children on a winning streak. And at university clubs throughout the world, old Blues profoundly wish them well.

Ballantine's Scotch was there. Like those classic days, the classic scotch. With a taste to be celebrated again and again.

Taste is why you buy it

Courtesy "21" Brands, Inc.

Appeal to Ignorance

The *appeal to ignorance* is an argument that uses an opponent's inability to disprove a conclusion as proof of the conclusion's correctness. The lesson the fallacy teaches, however, is that one cannot use another person's ignorance of the refutation of a thesis as proof of its confirmation. We cannot shift this burden of proof. We do this, of course, in the hope of putting our opponents on the defensive and thus causing them to believe the proposed conclusion must be true, seeing they cannot prove it is not. Exercises 535 to 549 are some typical examples of the fallacy; exercises 550 to 553 try to strengthen the appeal of this fallacy by combining it with complex question.

*535. Popular saying:

God exists because you cannot prove He doesn't.

536. Popular saying:

People have freedom of choice, for no one has been able to show that we do not.

537. Popular saying:

Mental telepathy is true because no one has been able to prove that it isn't.

*538. From a logic text:

People have no soul, for hundreds of doctors who have dissected every part of the human body have been unable to find one.

539. Overheard:

You can't prove he was to blame for the misfortune, so it must actually have been someone else who was responsible.

*540. Overheard:

You cannot name one thing this woman has done that would prove she is capable of the position; I cannot see, therefore, any grounds for voting for her.

541. Speaker:

No one has written to object to the peace terms. Therefore there is no reason whatever to believe the country is not solidly behind them.

*542. Overheard:

He did not make a mistake in his computation of his income tax, for I have never known him to make a mistake.

543. Speaker:

Two years ago I publicly declared she was guilty of falsifying her income tax return, and she has never sued me for perjury. Is this not an open admission of my charge?

544. Letter to the editor:

It can't be proven that the fetus isn't a real person because we don't know what the mental life of the fetus is like. So abortion is morally wrong.

545. From William Gerhardie:

"You believe in immortality?"
"I have not sufficient data not to believe in it."

William Gerhardie, *The Polyglots* (New York: Macdonald & Jane's Publishers Ltd., 1970), p. 252.

546. From Sheila Ostrander and Lynn Schroeder:

Said the Soviet Academy of Sciences in March 1968, "the search for UFOs is anti-scientific." If they existed, scientists would know about them. The Academy stated, "None of our astronomers have ever seen a UFO. They've never been sighted by any of our ground scientists. Our defensive units, guarding the land day and night, have never seen a UFO." So, the Academy concluded, there can't be any UFOs.

Sheila Ostrander and Lynn Schroeder, *Psychic Discoveries Behind the Iron Curtain* (New York: Bantam, 1971), p. 102. ©1970 by Sheila Ostrander and Lynn Schroeder. Published by Prentice-Hall, Inc. Englewood Cliffs, New Jersey 07632 and The Foley Agency.

547. Lionel Tiger, replying to criticism of his article, in which he argued that male dominance was genetically based:

(My critics) have the responsibility to provide a better scientific explanation for the observable data about sex differences in behavior before objecting to the one offered in my article.

New York Times, November 15, 1970. © 1970 by The New York Times Company. Reprinted by permission.

548. From *Roughing It* (1871):

(One man is telling another that a bull had chased him up a tree):
"... Sure enough, it was as I had dreaded; he started in to climb the tree—"
"What, the bull?"
"Of course—who else?"
"But a bull can't climb a tree."
"He can't, can't he? Since you know so much about it, did you ever see a bull try?"

Samuel Langhorne Clemens, *Roughing It* (New York: Harper & Row, Pub., 1913), p. 47.

549. News report:

On the Senate floor in 1950, Joe McCarthy announced that he had penetrated "Truman's iron curtain of secrecy." He had 81 case histories of persons who he considered to be Communists in the State Department. Of case 40, he said, "I do not have much information on this except the general statement of the agency that there is nothing in the files to disprove his Communist connections."

*550. Overheard:

Has anybody proved that Jones is not a Communist?

*551. Overheard:

What reason could anyone have for forging them?

552. Overheard:

If these writings are not Shakespeare's to whom do they belong?

553. Speaker:

How could the world have had no beginning?

Appeal to Fear

Appeal to fear is an argument, if one may call it that, in which one uses the threat of harm to advance one's conclusion. It goes under a variety of different names: *swinging the big stick; the scare technique;* and, in Latin, *argumentum ad baculum.* It is an attempt to influence people by threatening them with unpleasant consequences of some kind if they do not agree. It is effective because by arousing sufficient fear in persons it is frequently possible to make them believe things they would reject as false in calmer moments.

The remaining examples illustrate some of the forms this fallacy assumes—ranging from the out-and-out threat ("Your money or your life!") to those in which evidence of some kind is used in an effort to camouflage the threat.

554. **American Cancer Society advertisement:**

555.   Another ad from the American Cancer Society:

Mark Waters was a chain smoker. Wonder who'll get his office?

Too bad about Mark. Kept hearing the same thing everyone does about lung cancer. But, like so many people, he kept right on smoking cigarettes. Must have thought, "been smoking all my life... what good'll it do to stop now?" Fact is, once you've stopped smoking, no matter how long you've smoked, the body begins to reverse the damage done by cigarettes, provided cancer or emphysema have not developed. Next time you reach for a cigarette, think of Mark. Then think of your office—and your home.

American Cancer Society

556.   From *Tennis* magazine, April 1977:

While you're thinking about going to an All American Tennis Camp, someone else might be booking your room.

557. Interchange:

Seller: This one is better, but you can't afford it.
Buyer: That's the one I'll take.

*558. Advertisement:

Without Insurance,
Your Life
Would Be a Deathtrap.

*559. From Richard Grunberger:

The Nazis used to send the following notice to German readers who let their subscriptions lapse: "Our paper certainly deserves the support of every German. We shall continue to forward copies to you, and hope that you will not want to expose yourself to unfortunate consequences in the case of cancellation."

Richard Grunberger, *The 12-Year Reich: A Social History of Nazi Germany 1933–1945* (New York: Holt, Rinehart & Winston, 1971), p. 398. By permission of Holt, Rinehart and Winston and Harold Matson Company, Inc.

560. From "Sexism Strikes Out: New Jersey Girls Get to First Base in Little League." *Ms. Gazette News,* May 1974, p. 20:

In the city of Hoboken, New Jersey, last year, the Young Democrats Little League baseball team had a girl pitcher named Maria Pepe. Then the National Little League Office got wind of this transgressive behavior and warned the Hoboken team that its charter called for boys only. Either Maria was to be booted off the team, or the team would be booted out of the league.

*561. Speaker:

You should vote for Democratic candidates. If the Republicans retain control of Congress, we are sure to have a depression.

562. Former Governor George C. Wallace, during his 1968 campaign:

If any demonstrator ever lays down in front of my car, it'll be the last car he'll ever lay down in front of.

563.   Former Vice-President Spiro Agnew:

We can afford to separate them [militant dissidents] from our society with no more regret than we should feel over discarding rotten apples from a barrel.

564.

Get to know Conrac...

before you smell smoke

It's a good idea to get to know Conrac. Our Rixson-Firemark Smok-Chek™ door closers can help confine smoke and flame and automatically sound an alarm when fire threatens. Fire/life safety is only one of the growing electronic industries where Conrac is a basic manufacturer and marketer. If you know an architectural hardware consultant, ask him. He knows us well... and the better you know us, the better you'll like what we can do for you.

Courtesy Conrac Corporation Stamford, CT.

*565.   Letter to the editor:

One suggestion to those who don't care much for the police is that the next time they get in trouble, call a hippie!

*566.   News report:

Maj. Gen. John K. Singlaub told a House armed services sub-committee that Carter decided to remove the troops without asking the opinion of U.S. civilian and military representatives on the scene—who presumably would have recommended against it.

"I know of no U.S. or ROK (Republic of Korea) officer in any position of responsibility who agreed with the proposal to withdraw all U.S. ground force troops on the (five-year) schedule the president proposed," Singlaub said.

"What you are telling us is that every official, military and civilian, on the ground in Korea believes that the withdrawal of U.S. ground forces runs the risk of war," subcommittee Chairman Samuel S. Stratton (D.-N.Y.) said.

"That is absolutely correct." Singlaub replied.

567. Letter to the editor:

Until the politicians of both parties cease their ingrained habits of wheeling and dealing, I would suggest that all citizens adopt a new slogan to apply to any future elections. "KIO" would roll off the tongue smoothly and stand for "Kick Incumbents Out." This might cause a few legislators to think twice about the oil depletion allowances, the income tax breaks for the affluent, and stricter campaign contribution control.

568. Letter to the editor:

After reading your letters section almost daily, I've noticed a startling increase in hostility from your readers—hostility to politicians and big business foremost. I wonder if these powers that arbitrarily run our lives have noticed. If they continue to disregard this hostility from the people, they may have more trouble than they or their mercenary protectors can handle.

569. Letter to the editor:

Sandra Good, Squeaky Fromme's roommate, said, "We're going to start assassinating presidents and vice-presidents. . . ." Thereafter, Good was allowed to remain free. Do we have to wait until disaster strikes again?

*570. Letter to the editor:

Clearly, a commonsense compromise must be reached between the environmental ideal and the awesome reality of our nation's present energy shortage. Failure can only mean homes without heat, cars without gas, workers without jobs, and, we hasten to add, a movement without merit.

571. Speaker:

As a result of the Arab cutoff of our oil supply, I understand more clearly now the rationale of imperialism.

572. **Speaker:**

With millions of college grads seeking employment, it should not be hard to replace dissatisfied teachers who feel the urge to strike.

*573. **Letter to the editor:**

For those heroin addicts who are caught burglarizing and attempting to burglarize our homes, let us solve their problem by giving them what they want—heroin—in large quantities at a gulp—in other words, an overdose. That will end their miserable existence and reduce the threat to our homes, possessions, and possibly even our lives.

574. **Interchange:**

Child: I don't believe in Santa Claus.
Mother: You'd better, or he won't bring you any presents.

ANSWERS TO EXERCISES
PART THREE

Genetic Fallacy

434. Something is not identical with that from which it is derived. Assuming magic and animism to be nonsense, it does not follow that religion, because it derived or had its source in magic and animism, is likewise nonsense.

438. The popularity of psychoanalysis has made this sort of argument appear sounder than it really is. Instead of meeting opponents' arguments with evidence, we seek to psychoanalyze them. Here we might reply that a young, harmless, constitutional invalid may be capable of writing some very dangerous and powerful philosophy—as he indeed did.

Abusive ad Hominem

442. Instead of considering Mr. North's comments on their own merits, the arguer attacks Mr. North by calling him a "fascist" and "reactionary." The same is true of Mr. Stuart Chase, whom the arguer attacks as a "petty bourgeois American economist" and hence as being incapable of writing the truth.

446. An attempt to make the other person feel guilty, the argument fails to consider that an American can believe foreign soldiers to be the finest in the world or that foreigners can believe Americans make the finest soldiers in the world.

448. Instead of considering the bill to see if it should be supported by manufacturers, the speaker tries to imply the manufacturer is being inconsistent (and therefore dumb) in not doing so.

Circumstantial ad Hominem

452. Why should we allow ourselves to believe that because the people involved are conscientious objectors their only reason for defending their case is to escape the draft? Whether they have any sound arguments in favor of their case can be determined only by looking at the arguments themselves.

Tu Quoque

456. The fact that Turkey produces opium is not relevant to the question of whether or not Americans violated Turkish drug laws and have therefore been sent to prison.

459. Ira Levin is justifying his own scandalous writing by in effect saying to this magazine, "What about you? You slander people on the cover of your magazine so what right do you have to criticize my writing as being in bad taste?" This argument is irrelevant as to whether or not his writing is actually in bad taste.

460. The writer would like us to believe that an immoral person is incapable of giving moral advice or that such advice is worthless. People who live in glass houses should not throw stones is what he is in effect saying. But there is no reason why a stone thrown from a glass house cannot reach its mark.

465. Perhaps both are wrong in doing what they do, but the comparison (and there is a type of false analogy involved here) will not stand up to examination. Animals bred for consumption are slaughtered in as humane a way as possible and this is not the case in the so-called sport of hunting and trapping animals.

468. The fact that fish practice cannibalism is not a sufficient reason for our eating them. After all, members of certain African tribes practice cannibalism yet we do not use this fact as a justification for our eating human flesh. Because they do it does not mean that we should. (Benjamin Franklin realizes, as he lets us know toward the end of his remark, how easily we tend to rationalize when our interests are at stake.)

469. Just because the "Establishment" is violent does not make it right for the "Left" to be violent. The fact that the Establishment does it does not make it right.

473. The fact that the socialists (or anyone else for that matter) would also do it does not justify your doing it. If it is wrong, it does not matter who it is who has done it—it remains wrong.

Poisoning the Well

474. A remark like this leaves the other person without any possible defense. What can one possibly say, everyone now having been assured that whatever comes out of one's mouth is a lie?

476. This is a much too easy way out. If enough people do not seem to understand what you are saying, your way of saying it may be the reason (and not everyone's seeming stupidity).

478. Anyone who rejects the speaker's conclusion, we are told, is either an "insane bigot" or a "foul-mouthed slanderer." Thus regardless of how we respond, we condemn ourselves. If one is faced with such a dilemma, one ought to stand up and expose the tactic by saying, "At the risk of being declared an 'insane bigot' or a 'foul-mouthed slanderer' by that bigot and slanderer up there on stage, I affirm...."

479. This is more complicated than it first appears. It may be construed as an attempt on Senator Scott's part to accuse the *New Times* of having poisoned his well, for, as he affirms, to win he has to prove malice, which is very difficult. Should he lose, as is likely, he will succeed in confirming *New Times*'s allegation. But this could be construed as a case of his poisoning their well. If he wins the suit, *New Times* would be proven wrong; and if he loses the suit, he could claim that since malice must be proven and this is especially difficult, the fact that he has not succeeded is no proof that he is in fact the dumbest member of Congress, as they had alleged!

Mob Appeal

483. This is an enormously clever advertisement. It would be unseemly to "push" a product that has been and is the cause of so much private and public misery and so once again (as is so typical in liquor advertisements) the appeal is to our sense of patriotism and love of humanity.

489. This example is fuller and more interesting than number 483. There is more rabble rousing here and we are approaching an almost classic example of the mob appeal fallacy. As in most such appeals, there is an attempt to ignite people, to stir their deepest and most irrational passions, and to confuse and foment frenzy. The only thing missing here (it will more than be made up for in the remaining examples of mob appeal) is a good deal of screaming and shrieking.

491. Again the typical concoction: an attempt to substitute feelings for logical thinking by blasting away with explosive words ("war hungry," "unholy thirst," and "my sons' blood").

Appeal to Pity

495. The photograph of the elderly, homey couple standing in the doorway is used to kindle recollections of our own family and home and hence to make us want to fly home. We feel sorry for this elderly couple, whom we identify with our parents, and so come to feel we ought to fly home (on Boeing) and cheer them up.

499. A verdict of guilty or not guilty is determined strictly on the merits or the faults of the case, not whether the son is the sole support of his aged parents—which is no doubt unfortunate and sad but irrelevant.

503. Rather than being told how our donations will be used (and what percentage will actually reach those they are designed to help), an attempt is made to open our hearts and with it our pocketbooks by depicting the courage and determination of this crippled boy making his lonely way down the hard pavement. From a psychological point of view this is masterful, but it is logically objectionable and devious.

Appeal to Authority

504. Bankers, dentists, doctors, and so forth, while enjoying expertise in their respective fields, are hardly sources of authority in the intricacies of public education and its relation to government.

505. Three *physicists* taking *psychical* research seriously does not mean there must be something to it.

507. The mayor is not a person trained to detect such things and therefore his failure to find evidence of dishonesty is no proof there was none. (This could be explained as a case of appeal to ignorance as well.)

510. Science in Aristotle's day was not capable of determining the specific gravity of air or fire; so Aristotle, though a great philosopher, cannot be used as an authority on these matters.

514. The fact that everybody is wearing something does not mean it is good, ought to be worn, or I ought to wear it.

515. *Coyotes* (even ten million of them) are not authorities on what *we* ought to eat.

518. Because Boeing serves the most people does not necessarily mean it is the best. It may be the cheapest trijet or the most easily serviceable. Besides, how many people who fly a Boeing jet know who built it? Because passengers like an airplane's interior does not make it a good airplane.

528. This advertisement tries to appeal to the select few who try to balance frugality with pleasure. You save money driving to work yet buy a CB radio. Why not go all the way and stop drinking "ordinary" scotch and switch to Pinch 12-year-old scotch?

531. Prostitution is also as old as human history. Does that make it sacred as well?

533. This is an attempt to appeal to tradition and history for commercial gain. Our response should be that two hundred and eighty years of dealing with royalty is no reason we ought to buy the product or that it will be good for us.

Appeal to Ignorance

535. The fact that you cannot prove God does not exist is not proof He does.

538. Just because hundreds of doctors have been unable to find a soul (in a human body) does not mean there is no soul. The soul might not be "in" the body.

540. The inability to prove her capability does not mean she is not capable. It just shows you are unable to prove it.

542. You might be ignorant of the fact that he does make mistakes, and his tax computation might be one.

550. The tactic here is: you do not know, so Jones must be. But, as we should reply, just because no one has proved Jones is not a Communist is no reason to assume Jones is.

551. Just because you do not know any reason someone would forge them doesn't mean they might not in fact be forged. (The tactic here is again the same: you do not know they are forgeries, so they must be genuine.)

Appeal to Fear

558. This is an attempt to frighten us to buy more insurance than we probably need by using a fright-producing term like "death-trap."

559. Their subscribers no doubt got the message and quickly renewed their subscriptions. "Unfortunate consequences" is a nice little euphemism.

561. Rather than giving sound arguments for and against the candidates, this plays upon our fear of a depression to get us to vote for the candidate proposed.

565. This is an attempt to frighten those who do not like the police into liking them (or perhaps dissuade them from criticizing the police) by indicating what can happen if they did not have them and had to rely on hippies in an emergency.

566. What we require here is an argument, or some reasons, showing how a troop withdrawal would lead to war, not just the threat of war to frighten us from withdrawing them.

570. This is an attempt to force us into accepting the position advocated by listing the dire consequences (unsupported by argument) if we do not.

573. Instead of providing useful suggestions how we might better protect ourselves from heroin addicts, the writer simply threatens these addicts with death by overdose.

TEST ON THE FALLACIES
OF RELEVANCE

1. Letter to the editor:

It is very typical of plans advanced by the oil companies to deal with the energy crisis. You know who the plan is going to favor.

2.

EAT YOUR HEART OUT, RUSSIA.

Maybe Russia invented vodka. But it took Gilbey's American know-how to make vodka a lot better... to smooth it, to make it delightfully crisp and clean. Try Gilbey's— the vodka the Russians wish they'd invented.

GILBEY'S VODKA

You can't buy a better vodka for love nor rubles.

VODKA 80 PROOF DIST FROM 100% GRAIN W & A GILBEY LTD, CINN. D DISTR. BY NAT'L DIST PROD CO. PRODUCT OF U.S.A.

Courtesy National Distillers Product Company.

3. Speaker:

This birth control proposal is contrary to your religious principles; you should be the first one to reject, not support it.

4. From Richard Nixon's Checkers speech:

My family was one of modest circumstances and most of my early life was spent in a store out in East Whittier. It was a grocery store—one of those family enterprises. I worked my way through college and to a great extent through law school.

The only reason we were able to make it go was because my mother and dad have five boys and we all worked in the store.

Why do I feel so deeply? Why do I feel that in spite of the smears, the misunderstanding, the necessity for a man to come up here and bare his soul as I have? And I want to tell you why. Because you see, I love my country.

5. **Letter to the editor:**

Furthermore, I expect that with a little more imagination, Lewis and television reformers might set their sights higher. Let's reform and take the violence out of fairy tales—the witch in *Hansel and Gretel,* for example, kills and eats children. And the Bible should not be neglected, either.

6. **From a Ku Klux Klan circular:**

Every criminal, every gambler, every thug, every libertine, every girl ruiner, every home wrecker, every wife beater, every dope peddler, every moonshiner, every crooked politician, every pagan Papist priest, every shyster lawyer, every K. of C., every white slaver, every brothel madam, every Rome controlled newspaper, every black spider—is fighting the Klan. Think it over. Which side are you on?

7. **National Organization for Reformation of Marijuana Laws president in answer to "Is it true marijuana prices will double by next year?"**

NORML hasn't heard any confirmed reports on the subject so you can take this issue to be untrue.

8. **Advertisement:**

The man who drinks life rather than sips it. The man whose boldness is matched only by his good taste. The man whose demands upon his possessions are as uncompromising as his demands upon himself. For this special type of man, and for him only, the Stutz Blackhawk has been built.

9. **Statement by military officer:**

You seldom hear much about My Lai anymore except from the Communists.

10. **From Malcolm X's speech, "The Ballot or the Bullet":**

It's time for you and me to stop sitting in this country letting some cracker senators, northern senators and southern senators, sit in Washington, D.C., and come to a conclusion in their mind that you and I are supposed to have civil rights. There's no white man that's going to tell me anything about my rights....

In 1964 it's the ballot or the bullet.

ANSWERS TO TEST
ON THE FALLACIES
OF RELEVANCE

1. Circumstantial *ad hominem*
2. Mob appeal
3. Abusive *ad hominem*
4. Appeal to pity
5. *Tu quoque*
6. Genetic fallacy
7. Appeal to ignorance
8. Appeal to authority
9. Poisoning the well
10. Appeal to fear

The Disaster Lobby

The following is the text of the speech Mr. Shepard delivered Jan. 28, 1971 before the 44th Annual Meeting of the Soap and Detergent Association at the Waldorf-Astoria Hotel, New York City. Mr. Shepard is publisher of Look *magazine.*

One morning last fall I left my office here in New York and hailed a cab for Kennedy Airport. The driver had the radio tuned to one of those daytime talk shows where the participants take turns complaining about how terrible everything is.

Air pollution. Water pollution. Noise pollution. Racial unrest. Campus unrest. Overpopulation. Underemployment. You name it, they agonized over it. This went on all the way to Kennedy and as we pulled up at the terminal the driver turned to me and said—and I quote—"If things are all that bad, how come I feel so good?"

Ladies and gentlemen. I wonder how many Americans, pelted day after day by the voices of doom, ever ask themselves that question: "If things are all that bad, how come I feel so good?"

Well, I think I have the answer. We feel good because things *aren't* that bad. Today I would like to tell you how wrong the pessimists are, and to focus an overdue spotlight on the pessimists themselves. These are the people who, in the name of ecology or consumerism or some other "ology" or "ism," are laying siege to our state and federal governments, demanding laws to regulate industry on the premise that the United States is on the brink of catastrophe and only a brand-new socio-economic system can save us. I call these people The Disaster Lobby, and I regard them as the most dangerous men and women in America today.

Dangerous not only to the institutions they seek to destroy but to the consumers they are supposed to protect.

Let's begin with a close-in look at that drumbeat of despair I heard in the taxicab and that all of us hear almost every day. Just how much truth is there to the Diaster Lobby's complaints?

Take the one about the oxygen we breathe. The Disaster folks tell us that the burning of fuels by industry is using up the earth's oxygen and that, eventually, there won't be any left and we'll suffocate. False. The National Science Foundation recently collected air samples at 78 sites around the world and compared them with samples taken 61 years ago. Result? There is today precisely the same amount of oxygen in the air as there was in 1910—20.95 per cent.

But what about air pollution? You can't deny that our air is getting more fouled up all the time, says the Disaster Lobby. Wrong. I *can* deny it. Our air is getting less fouled up all the time, in city after city.

In New York City, for example, New York's Department of Air Resources reports a year-by-year *decrease* in air pollutants since 1965. What's more, the New York City air is immeasurably cleaner today than it was a hundred years ago, when people burned soft coal and you could cut the smog with a knife.

Which brings us to water pollution. The Disaster Lobby recalls that, back in the days before America was industrialized, our rivers and lakes were crystal clear. True, and those crystal-clear rivers and lakes were the source of the worst cholera, yellow fever and typhoid epidemics the world has ever known. Just one of these epidemics—in 1793—killed one of every five residents of Philadelphia.[1]*

Our waterways may not be as pretty as they used to be, but they aren't as deadly, either. In fact, the water we drink is the safest in the world. What's more, we're making progress cosmetically. Many of our streams will soon *look* as wholesome as they are.

Perhaps it's the fear of overpopulation that's getting you down. Well, cheer up. The birth rate in the United States has been dropping continuously since 1955 and is now at the lowest point in history. If the trend continues it is remotely possible that by the year 4,000 there won't be anyone left in the country. But I wouldn't fret about under-population either. Populations have a way of adjusting to conditions, and I have no doubt that our birth rate will pick up in due course.[2]

I now come to the case of the mercury in tuna fish. How did it get there? The Disaster Lobby says it came from American factories. The truth, as scientists will tell you, is that the mercury came from deposits in nature.

To attribute pollution of entire oceans to the 900 tons of mercury released into the environment each year by industry—that's less than 40 carloads—is like blaming a boy with a water pistol for the Johnstown Flood. Further proof? Fish caught 44 years ago and just

*The numbers indicate paragraphs containing a fallacy. The answers are in the *Teacher's Manual.*

analyzed contain twice as much mercury as any fish processed this year.[3]

Speaking of fish, what about the charge that our greed and carelessness are killing off species of animals? Well, it's true that about 50 species of wildlife will become extinct this century. But it's also true that 50 species became extinct *last* century. And the century before that. And the century before *that*.

In fact, says Dr. T.H. Jukes of the University of California, some 100 million species of animal life have become extinct since the world began. Animals come and animals go, as Mr. Darwin noted, and to blame ourselves for evolution would be the height of foolishness.[4]

Then there is the drug situation. Isn't it a fact that we are becoming a nation of addicts? No, it is not. Historically, we are becoming a nation of non-addicts. Seventy years ago, one of every 400 Americans was hooked on hard drugs. Today, it's one in 3,000. So, despite recent experimentation with drugs by teenagers, the long-range trend is downward, not upward.

Another crisis constructed of pure poppycock is the so-called "youth rebellion," to which the Disaster Lobby points with mingled alarm and glee. But once you examine the scene in depth—once you probe behind a very small gaggle of young trouble-makers who are sorely in need of an education, a spanking and a bath, not necessarily in that order—you can't find any rebellion worth talking about.[5]

A while back *Look* commissioned Gallup to do a study on the mood of America. Gallup found that, on virtually every issue, the views of teenagers coincided with those of adults. And on those issues where the kids did *not* see eye-to-eye with their elders, the youngsters often tended to be more conservative.

The same assessment can be made of the putative black rebellion. There isn't any. Oh, there are the rantings of a lunatic fringe—a few paranoid militants who in any other country would be behind bars and whose continued freedom here is testimony to the fact that we are the most liberated and least racist nation on earth. But the vast majority of black Americans, as that same Gallup study revealed, are staunch believers in this nation.

How about unemployment? The Disaster people regard it as a grave problem. Well, I suppose even one unemployed person is a grave problem, but the record book tells us that the current out-of-work level of 6 percent is about par. We've had less, but we've also had more— much more.

During the Kennedy Administration unemployment topped 7 percent. And back in the recovery period of Franklin Roosevelt's second term, unemployment reached 25 per cent. So let's not panic over this one.

That word "panic" brings me to the H-bomb. Some people have let the gloom-mongers scare them beyond rational response with talk about atomic annihilation. I can't guarantee immunity from the bomb, but I offer the following as food for thought. Since World War II, over one billion human beings who worried about A-bombs and H-bombs died of other causes. They worried for nothing. It's something to think about.[6]

One final comment on the subject. Members of the Disaster Lobby look back with fond nostalgia to the "good old days" when there weren't any nasty factories to pollute the air and kill the animals and drive people to distraction with misleading advertisements. But what was life *really* like in America 150 years ago?

For one thing, it was very brief. Life expectancy was 38 years for males. And it was a grueling 38 years. The work week was 72 hours. The average pay was $300. Per *year*, that is.

The women had it worse. Housewives worked 98 hours a week, and there wasn't a dishwasher or vacuum cleaner to be had. The food was monotonous and scarce. The clothes were rags. In the winter you froze and in summer you sweltered and when an epidemic came—and they came almost every year—it would probably carry off someone in your family. Chances are that in your entire lifetime you would never hear the sound of an orchestra or own a book or travel more than 20 miles from the place you were born.

Ladies and gentlemen, whatever American businessmen have done to bring us out of that paradise of 150 years ago, I say let's give them a grateful pat on the back—not a knife in it.

Now I'm not a Pollyanna. I am aware of the problems we face and of the need to find solutions and put them into effect. And I have nothing but praise for the many dedicated Americans who are devoting their lives to making this a better nation in a better world.

The point I am trying to make is that we are solving most of our problems, that conditions are getting better, not worse, that American Industry is spending over $3 billion a year to clean up the environment and additional billions to develop products that will keep it clean, and that the real danger today is not from the free enterprise establishment that has made ours the most prosperous, most powerful and most charitable nation on earth.

No, the danger today resides in the Disaster Lobby—those crepe-hangers who, for personal gain or out of sheer ignorance, are undermining the American system and threatening the lives and fortunes of the American people.

When I speak of a threat to lives, I mean it literally. A classic example of the dire things that can happen when the Disaster Lobby gets busy is the DDT story.

It begins during World War II when a safe, cheap and potent new insecticide made its debut. Known as DDT, it proved its value almost overnight. Grain fields once ravaged by insects began producing bumper crops. Marshland became habitable. And the death rate in many countries fell sharply.

According to the World Health Organization, malaria fatalities dropped from four million a year in the 1930s to less than a million by 1968. Other insect-borne diseases also loosened their grip. Encephalitis, Yellow fever, Typhus.[7]

Wherever DDT was used, the ailment abated. It has been estimated that a hundred million human beings who would have died of one of these afflictions are alive today because of DDT.

But that's not the whole story. In many countries famine was once a periodic visitor. Then, largely because of food surpluses made possible by DDT, famines became relatively rare. So you can credit this insecticide with saving additional hundreds of millions of lives.

Then in 1962 a lady named Rachel Carson wrote a book called *Silent Spring* in which she charged that DDT had killed some fish and some birds. That's all the Disaster Lobby needed.

It pounced on the book, embraced its claims—many of them still unsubstantiated—and ran off to Washington to demand a ban on DDT. And Washington meekly gave them their ban, in the form of a gradual DDT phase-out. Other countries followed the U.S. lead.

The effects were not long in coming. Malaria, virtually conquered throughout the world, is having a resurgence. Food production is down in many areas. And such pests as the gypsy moth, in hiding since the 1940s, are now munching away at American forests.

In some countries—among them Ceylon, Venezuela and Sweden—the renaissance of insects has been so devastating that laws against DDT have been repealed or amended. But in our country the use of DDT, down to 10 per cent of its former level, may soon be prohibited entirely.

The tragedy is that DDT, while it probably did kill a few birds and fish, never harmed a single human being except by accidental misuse. When the ultimate report is written, it may show that the opponents of DDT—despite the best of intentions—contributed to the deaths of more human beings than did all of the natural disasters in history.

In addition to endangering human life, the Diaster Lobbyists are making things as difficult as possible for us survivors. By preventing electric companies from building new power plants, they have caused most of those blackouts we've been experiencing.

By winning the fight for compulsory seat belts in automobiles, they have forced the 67 per cent of all Americans who do not use seat belts to waste $250 million a year buying them anyway.[8]

By demanding fewer sizes in package goods on the ground that this will make shopping easier for the handful of dumbbells in our society, they are preventing the intelligent majority of housewives from buying merchandise in the quantities most convenient and most efficient for their needs.[9]

And I don't have to tell anyone in this room what the Disaster crowd has done and is doing to make washday a nightmare in millions of American homes. By having the sale of detergents banned in some areas and by stirring up needless fears throughout the country, they have missed the point entirely.

As Vice President Charles Bueltman of the Soap and Detergent Association recently pointed out, detergents with phosphates are perfectly safe, eminently effective and admirably cheap. And if they foam up the water supply in some communities, the obvious remedy is an improved sewer system. To ban detergents is the kind of overkill that might be compared with burning down your house to get rid of termites.

But of all activities of the Disaster Lobbyists, the most insidious are their attempts to destroy our free enterprise system. And they are succeeding only too well.

According to Prof. Yale Brozen of the University of Chicago, free enterprise in the United States is only half alive. He cited as evidence our government's control of the mail, water supplies, schools, airlines, railroads, highways, banks, farms, utilities and insurance companies, along with its regulatory involvement in other industries.

And his statement was made prior to introduction in Congress last year of 150 bills designed to broaden government influence over private business.

Fortunately, most of the bills were defeated or died in committee. But they will be back in the hopper this year, along with some new bills. And they will have support from the darlings of the Disaster Lobby—senators like Moss, Proxmire and Hart and Representatives like Rosenthal of New York.

If so many important people are against free enterprise, is it worth saving? I think it is. With all its faults, it is by far the best system yet devised for the production, distribution and widespread enjoyment of goods and services.

It is more than coincidence that virtually all of mankind's scientific progress came in the two centuries when free enterprise was operative in the Western world, and that most of the progress was achieved in the nation regarded as the leading exponent of free enterprise: the United States of America.

For in the past 200 years—an eyeblink in history—an America geared to private industry has conquered communicable diseases, abolished starvation, brought literacy to the masses, transported men to another planet and expanded the horizons of its citizens to an almost incredible degree by giving them wheels and wings and electronic extensions of their eyes, their ears, their hands, even their brains. It has made available to the average American luxuries that a short time ago were beyond the reach of the wealthiest plutocrat.[10]

And by developing quick-cook meals and labor-saving appliances, it has cut kitchen chores in most homes from five hours a day to an hour and a half—and as a result has done more to liberate women than all of the bra-burning Betty Friedans, Gloria Steinems and Kate Milletts combined.

But the practical benefits of free enterprise are not my principal reason for wanting to preserve the system. To me the chief advantage of free enterprise is in the word "free." "Free" as opposed to controlled. "Free" as opposed to repressed. "Free" as in "freedom."

I am always amazed that members of the Disaster Lobby— libertarians who champion the cause of freedom from every podium, who insist on everyone's right to dissent, to demonstrate, to curse policemen and smoke pot and burn draft cards and fly the flags of our enemies while trampling our own—these jealous guardians of every citizen's prerogative to act and speak without government restraint are also the most outspoken advocates of eliminating freedom in one area. When it comes to commerce, to the making and marketing of goods, our liberty-loving Disaster Lobby is in favor of replacing freedom with rigid controls.

And let us not minimize the value of this freedom of commerce to every man, woman and child in our country.

This is the freedom that makes it possible for the consumer to buy one quart of milk at a time—even though a government economist may think gallon containers are more efficient and quarts should be abolished.

This is the freedom that enables the consumer to buy rye bread if he prefers the taste—although someone in Washington may feel that whole-wheat is more nutritious and rye should be outlawed.

This is the freedom that allows the consumer to buy a refrigerator in avocado green despite some bureaucrat's desire to have all refrigerators made in white because it would be more economical that way.

For in a free economy, the consumer—through his pocketbook— determines what is made and what is sold. The consumer dictates the sizes, the shapes, the quality, the color, even the price.

And anyone who doubts the importance of this element of freedom ought to visit one of those grim, drab countries where the government decides what should and what should not be marketed.

But this is the direction in which the Disaster Lobby is pushing our country. What surprises me is how few of us seem to recognize the enormity of the threat. Instead of fighting back, we keep giving in to each insane demand of the consumerists—in the hope, I suppose, that if we are accommodating enough the danger will go away.

Well, ladies and gentlemen, it won't go away. If I accomplish nothing else today, I hope I can make that fact transparently plain.

Take the Nader group, for example, I have heard many businessmen dismiss Ralph Nader and his associates as well-meaning fellows who sincerely want to help the consumer by improving business methods.

Forget it. Mr. Nader isn't interested in seeing American industry clean house. What he wants is the *house*—from cellar to attic. His goal is a top-to-bottom takeover of industry by the government, with Mr. Nader, himself, I would guess, in charge of the appropriate commission.

Find it hard to believe? Then listen to this Associated Press report of a speech he made last September:

"Consumer advocate Ralph Nader has proposed that corporations that abuse the public interest should be transferred to public trusteeship and their officers sent to jail."[11]

Well, we all know which corporations abused the public interest in the eyes of Mr. Nader, don't we. *All* of them. The automobile companies. The tire companies. The appliance companies. The drug companies. The food companies. And yes, indeed, the soap and detergent companies.

What Mr. Nader really desires, ladies and gentlemen, is for the government to take over *your* companies and to toss all of you into the calaboose, presumably without a trial. At least he never said anything about a trial.

Does anyone still think Mr. Nader and the rest of the Disaster Lobby are just some harmless do-gooders? Those who know them best don't think so. Federal Trade Commissioner Paul Rand Dixon, for example. Not long ago, he said of Mr.Nader—and I quote—"He's preaching revolution, and I'm scared."

So let's start fighting back! It's not an impossible task because the Disaster Lobby is, by and large, not too bright and far too preposterous. All we have to do to win over the American people is acquaint them with the facts.[12]

We must show them that the consumerists are for the most part devout snobs who believe that the average man is too stupid to make his own selections in a free marketplace.

Our Disaster group opponents also have the most cockeyed set of priorities I have ever encountered. To save a few trees, they would prevent construction of a power plant that could provide essential electricity to scores of hospitals and schools. To protect some birds, they would deprive mankind of food. To keep fish healthy, they would allow human beings to become sick.[13]

One curious feature of the Disaster Lobby is an almost total lack of ethics. I say "curious" because these are the people who demand the maximum in ethics from private industry.[14]

Not long ago, an organization favoring clean air ran an ad soliciting funds from New Yorkers. It was full of halftruths and nontruths, including this sentence: "The longer you live with New York's polluted air and the worse it gets, the better your chances of dying from it." But we know that New York's air is *not* getting worse. Just let some private company run that ad and see how fast the consumerism boys would have a complaint on file with the FTC.

Immaturity is also a characteristic of the Disaster man. His favorite question is, Why can't we have everything? Why can't we have simonpure air and plentiful electricity *and* low utility rates, all at the same time? Why can't we have ample food *and* a ban on pesticides? I recommend the same answer you would give a not-too-intelligent five-year-old who asks, "Why can't I eat that cookie and still have it?" You explain that you just *can't* under our present technology.

Just recently the Coca-Cola Co. felt it necessary to reply to environmentalists who demand immediate replacement of glass and metal soft drink containers with something that will self-destruct. "A degradable soft drink container sounds like a fine idea," said Coca-Cola, "but it doesn't exist. And the chances are that one can't be made."

And Edward Cole, president of General Motors, responding to a government mandate for drastic reductions in exhaust emission within the next four years, stated: "The technology does not exist at this time—inside or outside the automobile industry—to meet these stringent emission levels in the specified time."

This inability of the Disaster people to accept reality is reflected in their frequent complaint that mankind interferes with nature. Such a thing is patently impossible. Man is *part* of nature. We didn't come here from some other planet. Anything we do, we do as card carrying instruments of nature.

You don't accuse a beaver of interferring with nature when it chops down a tree to build a dam. Then why condemn human beings for chopping down a lot of trees to build a lot of dams—or to do anything else that will make their lives safer or longer or more enjoyable.[15]

When it comes to a choice between saving human lives and saving some fish. I will sacrifice the fish without a whimper. It's not that I'm anti-fish; it's just that I am pro-people.[16]

The Disaster Lobbyist's immaturity shows up again and again in his unwillingness to compromise, to understand that man must settle for less than perfection, for less than zero risk, if he is to flourish. Failing to understand, they demand what they call "adequate testing" before any new product is released to the public. But what they mean by adequate testing would, if carried out, destroy all progress.

If penicillin had been tested the way the Disaster Lobby want all products tested—not only on the current generation but on future generations to determine hereditary effects—this wonder drug would not be in use today. And millions of people whose lives have been saved by penicillin would be dead.

We simply cannot test every aspect of human endeavor, generation after generation, to make *absolutely* certain that *everything* we do is *totally* guaranteed not to harm *anybody* to *any* degree whatsoever. We must take an occasional risk to do the greater good for the greater number. But that is a rational, mature evaluation—something of which the Disaster Lobby seems utterly incapable.

So this is the face of the enemy. Not a very impressive face. Not even a pleasant face. We have nothing to lose, therefore, by exposing it to the American people for what it is.

The time for surrender and accommodation is past. We must let the American public know that once free enterprise succumbs to the attacks of the consumerists and the ecologists and the rest of the Disaster Lobby, the freedom of the consumer goes with it. His freedom to live the way he wants and to buy the things he wants without some Big Brother in Washington telling him he can't.

Truth and justice and common sense are on our side. And Americans have a history of responding to these arguments. All we have to do is get the story out—as often as possible in as many forms as possible. And let's not vitiate our efforts by talking to each other—one businessman to a fellow businessman.[17]

The people we must reach are the *consumers* of America, and they're out there right now listening to propaganda from the others side and as often as not agreeing with it. But why shouldn't they? They have yet to hear the truth.

It's a bit late to make a New Year's resolution, but I suggest this one for anyone willing to chip in with a tardy entry. Let us resolve that 1971 will be the year we help convince the people of America that our nation is a great one, that our future is a bright one and that the Disaster Lobby is precisely what the name implies. A disaster.[18]